the one hand, he explores the influence of the concept of death on the current effort to decelerate aging and extend life span. On the other, he presents the socio-ethical issues related to human enhancement to lay audiences. He is also a columnist for the *Huffington Post* (France) as well as co-founder, vice-president and senior editor of NeoHumanitas, a Swiss think-tank, whose objectives are to democratize discussions on the socio-ethical issues related to the modifications of individuals through technological interventions.

Ruud ter Meulen is a psychologist and an ethicist. He is Chair for Ethics of Medicine and Director of the Centre for Ethics in Medicine at the University of Bristol. Previously he worked as Professor of Philosophy and Director of the Institute for Bioethics at the University of Maastricht. He has been working on a range of issues in medical ethics, including issues of solidarity and justice in health care, ethics of human enhancement, ethical issues of health care reform and health policy, ethics of evidence-based medicine, ethical issues of long-term care and ethics of research. He was coordinator of the European ENHANCE project on the ethical, philosophical and social issues of enhancement technologies, the EPOCH project on the role of ethics in public policy-making on new biotechnologies, and the SYBHEL project on the ethical, legal and social issues of synthetic biology as applied to human health.

Jean-Noël Missa is Research Director at the Belgian National Fund for Scientific Research and a professor at the Université Libre de Bruxelles (ULB). Trained in both medicine and philosophy, he was Fulbright Visiting Research Scholar at New York University (NYU) from 2002 to 2003. He is Director of the Center for Interdisciplinary Research in Bioethics (CRIB) and a member of the Belgian Advisory Committee on Bioethics. Most of his work concerns the philosophy and ethics of biomedicine, particularly in the field of neuroscience and biological psychiatry. He has published widely on these topics. He is a member of the Royal Academies for Science and the Arts of Belgium. He has also been appointed as a visiting professor at several universities (Paris, Abidjan, Bogotá).

Pascal Nouvel is Professor of Philosophy at the University Paul-Valéry Montpellier 3 and Head of the Centre d'Ethique Contemporaine (Centre for Contemporary Ethics), a research team at that university. He has a PhD in Science (Biology) and in Philosophy. His research deals mainly with philosophy of the life sciences and more particularly with the sciences of affects, emotions and passions. He was formerly Programme

Director at the Collège International de Philosophie in Paris where he led a programme on the 'Epistemology of Affects and Feelings' that aimed to investigate the kind of knowledge contained in our affects. As a part of this research programme, he took a special interest in investigating the affect-inducing power of certain psychotropic drugs, mainly amphetamines. This research led to the publication of two books: *Histoire des amphétamines* (2009) and *Axiomatique des sentiments* (2015).

Brian Stableford has been a writer all his life. He has published more than 70 novels and 30 short story collections, in addition to 30 non-fiction books and more than 150 volumes of translations from the French. He is currently researching a book on the history of the French *roman scientifique* from 1700 to 1940, supplemented by approximately 150 volumes in English of previously untranslated works in that genre – a project that he hopes to complete in 2016.

Myriam Winance is a research scholar with the French National Institute for Health and Medical Research (INSERM), working at the Centre for Research on Medicine, Science, Health and Society (CERMES3). Her research goal is to analyse the evolution of the notion of disability through discourses and processes of care (on a macro level) and the concrete practices and experience of disability (on a micro level). She is interested in the changes in the relation between 'disability' and the notion of 'repair', as reflected over time in French disability policies. She explores the way disability has been defined throughout history through successive 'repair' schemes and the way those schemes shape the 'disabled' person and his/her rights. Concerning the analysis of concrete practices, she has launched an ethnographic and interview-based research project on the wheelchair, examining how using a wheelchair transforms a person, and how norms of action and relationship are mobilised, questioned and reinterpreted.

Introduction

Simone Bateman, Jean Gayon, Sylvie Allouche, Jérôme Goffette and Michela Marzano

What is human enhancement? This term has become so common that it hardly seems to need clarification. And yet this is precisely the issue we raise in this book: the definition, scope and limits of human enhancement are as vague as the term is salient. Books explicitly devoted to human enhancement have already been published, mainly by philosophers, but most of these books deal with its ethical and political aspects, such as: Is it good and desirable or bad and dangerous? What moral criteria can help us shape our judgements? A selection of such books are listed at the end of the introduction; many are referred to throughout this book. Less attention, however, is devoted to questions about what exactly is being studied and how it should be approached.

What is human enhancement all about? What does 'the improvement of human capacities' mean? Is it relevant to distinguish current practices meant to restore health, compensate disabilities or improve appearance from enhancement practices? What practices today could be labelled as actual or potential forms of human enhancement? Are they similar to the enhancement practices envisioned by global projects or utopias such as transhumanism?

Human enhancement has become an issue of primary concern in debates about the future of contemporary societies. Yet this book does not argue for or against human enhancement, nor does it propose a detailed discussion of the ethical problems that it raises. Normative issues are not ignored, but they are not the *primum movens* of the inquiry. At the origin of this critical inquiry into human enhancement is the ambiguity of the term and the uncertain limits of its semantic domain. There is indeed no consensual definition of human enhancement: in this term, both 'human' and 'enhancement', that is, both the target and the action, are understood differently by different protagonists. How is the term 'human' understood? As human capacities, human individuals, the human condition or possibly the human species? And, even if the target is settled, is the term 'enhancement' a synonym for 'improvement', or does it mean 'augmentation', as suggested by the translation that seems to prevail now in Latin languages? Furthermore, 'enhancement' is both a descriptive term and a value-laden term, not to speak of those who consider it a normative term (that is, enhancement understood as an obligation). The semantic field of 'human enhancement' turns out to be a colourful panoply of heterogeneous practices – either actual or imagined – that generates controversies, aspirations, fears and dreams. In other words, the term is put to use in many different contexts, and is subject to unending interpretations, most of which are a justification of some end. Deciding which meaning of the term 'human enhancement' will prevail, with its respective practices, goals and justifications, will necessarily be the object of fierce competition.

Intrigued by this situation, we felt it was necessary to call on contributors from different countries and from different disciplinary backgrounds – sociology, philosophy, bioethics, political science, engineering, medicine, literary studies and also a renowned science fiction writer – to be able to comprehend human enhancement in its multiple dimensions. By doing so, we hope to have opened new routes for investigation, rather than refining pre-existing descriptions and arguments about human enhancement.

The topics examined by the authors address three fundamental aspects of human enhancement: what it means (concept), what it is (practices) and what it might be (visions of the future).

Human enhancement: what do we mean?

A common feature of many concrete practices identified today as forms of human enhancement, such as off-label prescription of growth hormone or blood-doping in sports, is that they are often proposed in the context of medical care and are supported by scientific and technical progress in biology and medicine. Understandably, many scholars have therefore attempted to understand human enhancement from the perspective of what distinguishes these practices from standard medical care, usually referred to as the 'therapy-enhancement distinction'. However, a number of enhancement practices are now developing in arenas increasingly disconnected from the medical world, making this approach too narrow to provide a full view of what human enhancement might be.

The first part of the book therefore proposes novel perspectives for exploring the meaning of human enhancement, with special regard to the set of practices, goals and justifications that are implied by this term. Each chapter proposes a unique point of view from which the field of human enhancement can be comprehensively understood. One author concentrates on subjective motivations to be enhanced and their underlying social determinants (Menuz); another prefers to focus on the specific characteristics of enhancement practices (Goffette), while a third (ter Meulen) proposes a comprehensive ethical evaluation of enhancement technologies. Bateman and Gayon adopt a general framework to examine the semantics of the term 'human enhancement' in its historical and social dimensions.

In their introductory essay, Simone Bateman and Jean Gayon explore the meaning – or rather meanings – of human enhancement while keeping in mind the question 'What is at stake?' They observe that 'human enhancement' – precisely that term and not other ordinary or scientific uses of the term enhancement – emerged in the mid-1990s, becoming widespread in the early 2000s. The spectacular dissemination of the term benefited, in their view, from the superposition of three different uses, most often not clearly distinguished, but conceptually distinct: improvement of human capacities, self-improvement and improvement of human nature. The enhancement of capacities seems to have emerged as a first usage in the context of bioethics, as an expansion of previous intense debates over the *genetic* enhancement of humans. In 2002, the publication of the National Science Foundation report, 'Converging technologies for improving human performance', gave a decisive impulse to this trend, because it introduced a new set of technologies, which extended the vision of what human enhancement might be far beyond the medical arena. The second layer of meaning, self-improvement, entails the idea that enhancement technologies may also serve a personal quest for identity and authenticity. Self-improvement is not so much an explicit doctrine as it is a subjective posture regarding the desirability of enhancement. The aspiration to improve oneself is in fact so deeply rooted in contemporary culture that it could become an important driving force in the promotion of these technologies. In contrast, the third semantic layer, improvement of human nature, refers to explicit doctrines claiming that modern technologies make it possible for humans to radically alter their capacities, improve their condition and even, according to some authors (especially the transhumanists), transform the human species. Although these doctrines have roots in history, among them eugenics and the ideologies of progress, it is the transhumanists who have been most active in promoting human enhancement as a vision of the future. Whether or not radical enhancements will be perceived by individual members of society as desirable, 'liveable' improvements of the *self* is likely to become a crucial arbitrating point in the debate.

Although acknowledging the semantic imprecision of the word 'enhancement', Jérôme Goffette claims that the term nonetheless encompasses a homogeneous set of practices, which he divides into eight categories: performance doping, use of psychostimulants (apart from medical indications), aesthetic transformations (apart from reconstructive plastic surgery), reproductive control (except in cases of reproductive dysfunction), mood modification (apart from medical indications), sex reassignment, quest for youth and immortality, and fabrication of a human being. All of these activities imply a logic of modification and improvement through the mastery of the body that goes beyond standard health care. Although enhancement and medicine may share similar resources, their respective goals are decidedly different. Hence his bold proposal: a new kind of activity is emerging, the art of 'changing and improving one's being by bodily modification'. Goffette proposes a name for this new art, 'anthropotechnics', and urges modern societies to recognise it as a full profession, with specific competencies, methods, and legal and deontological rules. In contrast with the therapy-enhancement distinction, which involves many semantic ambiguities, Goffette argues that the distinction between medicine and anthropotechnics is much clearer, and *should* be sharply made, both for conceptual and practical reasons. His focus on actual practices points to the first layer of meaning that Bateman and Gayon identify as improvement of human capacities.

Another way of encompassing the field of human enhancement is put forward by Vincent Menuz, who proposes a typology of enhancement practices based on individual motivations to be enhanced. Menuz gives an impressive table of these motivations, which he classifies into four overarching categories: (1) interventions aimed at adapting to the environment (skills in many areas such as work, sport, fighting, sexual attraction, and so on); (2) interventions aimed at fighting disease, aging and death; (3) interventions on existing or future children; and (4) interventions aimed at increasing happiness/well-being. As acknowledged by the author, this classification resembles that used by the US President's Council of Bioethics in their 2003 report on human enhancement practices entitled *Beyond Therapy*: 'Superior performance', 'Ageless bodies', 'Better children', and 'Happy souls'. But Menuz' classification is obtained in a very different way: it is inductively extracted from an extensive array of publications on enhancement. The table provided shows that the first category, which boils down to improving performance, overwhelms all the others by the number of its sub-categories. Menuz insists that his classification is based on personal motivations: 'Prior to any biotechnological intervention, there is an appeal for it'. But the main point of the chapter is that the personal or subjective motivations to be enhanced all result from strong social pressures. These pressures are particularly important in forms of enhancement that aim to improve performance and social adaptation. Menuz' analysis confirms the importance of Bateman and Gayon's second layer of meaning of enhancement, 'self-improvement', delicately balanced between subjective perceptions and socio-cultural forces.

In his assessment of the ethical issues involved in human enhancement, Ruud ter Meulen offers a comprehensive view of human enhancement (practices and discourses) in terms of the kinds of moral concerns that they generate. He points to three major ethical challenges commonly discussed in the literature: a decrease in solidarity, a reinforcement of individualism and a weakening of individual responsibility. The first challenge might result from an unequal distribution of enhanced capacities, thus creating divisions within a society: people without enhanced capacities would be at risk of being seen as inferior and possessing fewer human rights. The second challenge is a possible loss of the sense of community with others, derived from a culture obsessed by self-improvement. The final challenge is that technological enhancement might lead people to replace individual agency and responsibility by a slavish use of drugs and other technologies to enhance their capacities. Even if these moral challenges are real, ter Meulen emphasises that these pessimistic moral scenarios are in no way inevitable. Enhancement technologies might also improve sociability (if equally shared) and generate new forms of authenticity and dignity. This will depend on the kind of society that individuals will be able to build together. Therefore, ter Meulen's attitude towards enhancement technology resembles the classical neutralist argument: technological resources are neither intrinsically good nor bad, they are what human collectivities decide to do with them.

Learning from enhancement practices

The second part of the book focuses on current enhancement practices. Its purpose is not to provide an extensive account of these practices, but rather to concentrate on a sample of actual enhancement technologies, and draw lessons from them. Several authors in the book (Eskiizmirliler, Goffette, Menuz, Missa) insist that human enhancement should not be seen exclusively or primarily as speculations about the future, but as resources and activities that are presently at our disposal. Some are well-known and widely diffused, such as doping and other technical aids for athletes. Others, such as Brain–Machine Interface (BMI) technologies, are just beginning to be applied in humans, usually in an experimental context, but they are already in a state of scientific development that allows realistic predictions about their potential for enhancement in a not-too-distant future. These case studies raise several problems that deserve close attention. Does it make sense to assess the degree of artificiality of an enhancement technology, knowing that efficient improvement of the human body often means complete incorporation of a change? How important is the process of individual and social adjustment to a given technology in attaining an improved or enhanced state? What are the social factors that lead individuals to make use of any given enhancement technology (such as doping)? The following four chapters provide some insight into these questions.

Pascal Nouvel, a biologist and philosopher, takes as his case study of human enhancement the use by athletes of erythropoietin, a hormone that controls red blood cell production. From this perspective, he challenges the relevance of the natural-artificial distinction in understanding enhancement and proposes instead a different approach. Nouvel classifies enhancement technologies according to two criteria: space and time. Is the effect of the substance, device or technological intervention localised or delocalised in the body? Is it temporary or definitive? Thus a removable prosthesis (for example, Pistorius' running blades) is localised and temporary; a doping substance (such as erythropoietin) is delocalised but temporary; a somatic genetic modification is localised and durable, possibly definitive; a germline modification is delocalised and definitive. Space and time thus provide a powerful and original way of classifying enhancement interventions, especially when they consist of clearly identifiable material entities introduced into the body. Nouvel's approach has two interesting consequences. First, it shows that the natural-artificial distinction is easily blurred when applied to enhancement. Such is the case of the ski champion Eero Mäntyranta who had inherited a mutated gene that naturally increased the rate of synthesis of erythropoietin. The same result can be temporarily obtained by injecting erythropoietin into an athlete's body, by inserting the appropriate gene in the zygote of a future child, or by selecting for the gene in successive generations. This thought experiment leads Nouvel to a second and paradoxical observation: the most radical transformation (delocalised and durable) is precisely that which mimics a '"natural" transformation'. His conclusions counter the common fantasies we have about what radically enhanced humans might look like. Although Nouvel does not draw ethical lessons from his analysis, his classification of enhancement practices in terms of space and time highlights not only the magnitude of a given modification, but also its possible irreversibility – which should be a matter of concern for an ethicist.

Time is also an important matter for Myriam Winance, Anne Marcellini and Eric de Léséleuc, but from a totally different perspective. Social scientists working on disability and sport, these authors raise the question of the time and effort it takes for someone – whether disabled or able-bodied – to adapt herself to a technical aid. From that point of view, compensation of disability and enhancement are one and the same thing. Winance et al. make a suggestive comparison between persons who use wheelchairs and paralympic athletes, such as Pistorius, who use running blades. In both cases, a new functional capacity is acquired. The wheelchair compensates (rather than repairs) a deficiency and therefore restores a capacity to move. Similarly Pistorius' prostheses compensate for the loss of his natural legs and restore his capacity to move; however, his running blades are designed specifically for use in athletic events. In both cases the technical aid reduces functional inequality but the goals are different: in the first case it is integration into ordinary life, whereas in the second, it is competitive performance. Nonetheless, in both cases, the use of a technical aid involves an invisible process of adjustment to the technical aid (with physiological, psychological and social dimensions). These adaptive processes, including arrangements made with one's social environment to make a technical aid effective, are for the most part invisible, creating the illusion that humans can be enhanced without effort through technology. Winance et al. call this the 'biotechnological illusion'. This kind of analysis could be generalised and applied to a number of technological enhancements. Even doping, the effects of which seem so immediate and unintentional, requires, in the long run, a delicate adjustment that, in practice, relies both on personal experience and on professional competencies that go far beyond the athlete alone.

The next two chapters focus on two distinct kinds of technical aids that involve massive investments and research: brain–machine interfaces and doping. BMI is probably one of the fastest growing areas of research; so far it has few applications, which are mainly experimental, but the prospects are considerable and might well entail the development of enhancements affecting numerous aspects of professional and private life, including our relations with other people. Doping, in contrast, is today probably the most frequently encountered example of enhancement, the effects of which are well-known and widely described and discussed. We have chosen these two examples because they both provide realistic case studies of enhancement as it exists today and as it might develop in the future.

The chapter on BMI is the result of a collaboration between an engineer working in computational neuroscience, Selim Eskiizmirliler, and a philosopher, Jérôme Goffette. The authors first propose a precise characterisation of what BMI technologies really are. A BMI system is a combination of software and hardware which translates neural activity into motor commands that trigger and control the operations of a machine. For the moment, brain–machine interfaces are only open loop systems, with no sensory feedback, but this limitation will probably be overcome in the near future. The best-known example is that of a monkey feeding itself with a robotic arm which it controls by means of the interface that links the arm to its brain. Similar experimental set-ups have involved persons with a disability, illustrated by a recent video in which a paraplegic woman intentionally moves a robotic arm so that she can drink from a covered cup. Eskiizmirliler and Goffette propose an impressive list of future uses of BMI systems affecting innumerable aspects of our daily lives: the acquisition of additional functional, albeit artificial, arms; the possibility of functioning on a scale larger than our body; the development of cross-human or collaborative BMI prostheses, allowing for instance a surgeon to control the acts and emotions of another surgeon at a distance. These are just a few examples to which, one imagines, the adjustment process described in the preceding chapter by Winance *et al.* might apply. The technological and economic perspectives, including their industrial and military uses, are huge, but so are the ethical problems that may arise from such uses. The authors insist that such technology generates possibilities that would certainly favour greater freedom of movement, action and interaction; but because it modifies the ordinary conditions of body ownership, it might also entail new forms of alienation, especially if an enhanced body is partly owned by a firm, for either medical or professional reasons. BMI thus offers a nice example of an enhancement technology that generates ethical ambivalence, as emphasised by Ruud ter Meulen in Chapter 4.

The chapter on doping behaviour is also the result of a collaboration of a practicing specialist, Patrick Laure (epidemiology and prevention of doping behaviour) and a philosopher, Sylvie Allouche. The authors initially focus on doping in sport. This behaviour concerns a very limited fraction of the population but it offers a remarkably well-documented case. The chapter provides statistics concerning adults and young people: for all sports, doping affects from 5 to 15% of adults, and significantly less among younger people. Statistics also show that the major motivation for doping is competition and that its prevalence is mainly a function of the level of competition; prevalence is also higher among athletes who are most committed to their sports. Therefore, Laure and Allouche conclude that performance pressure is the 'prime mover' of doping behaviour, even if they also take into consideration the possibility that taking drugs may increase the pressure to perform. Interestingly, the authors ask whether this conclusion can be extended to doping behaviour outside sport. Data are not as abundant and as well-established in these domains. As in sport, personal involvement in an activity that is highly competitive is a key factor in doping behaviour. However, in contrast to sport, pressure to perform is less dependent on one's hierarchy in a profession; furthermore, professions that are vulnerable to fatigue or stress are more exposed to the risk of doping behaviour. Can this result be extended to other, and perhaps all, forms of technological enhancement? The case of doping, because it is so well documented, provides an important reference.

Visions of the future: lessons from art and fiction

Imagination plays a major role in current representations of and reflection on human enhancement. Indeed, human enhancement does not simply modify an individual's capacities or performance; it simultaneously changes the way people relate to one another. Questioning the concept and the practices of human enhancement ultimately raises questions about the kind of society in which we wish to live tomorrow and the values we wish to promote, thus inviting us to explore the imaginary constructs disseminated by contemporary culture. Literature, art and, more particularly, science fiction have provided exceptional means for imagining social relationships in a world where human enhancement is a common practice. Scientists themselves have both influenced and made important contributions to this anticipatory literature. The third part of the book is devoted to examining these scenarios, both as cultural phenomena and as large-scale thought experiments that nourish prospective thinking.

The philosophies of progress elaborated in the 17th and 18th centuries (for example, by Francis Bacon and Nicolas de Condorcet) are often mentioned as remote ancestors of contemporary discourses on human enhancement. Indeed, they frequently insisted on the capacity of science and technology not only to give humankind the power to master the world, but also to modify human beings themselves, their bodies and their minds. Christopher Coenen points to more direct antecedents. Coenen gives detailed evidence of authors who, soon after Darwin, began to defend the idea that science alone, rather than political reforms, could radically improve the human condition, formulating the same kind of promises as today's transhumanists (such as eradication of disease, self-directed evolution, human–machine hybrids, life extended to hundreds of years, unlimited material progress, and so on). In contrast with contemporary authors who try to find in their 'precursors' a legitimation of their present discourse, this historical reconstruction identifies a precise intellectual genealogy, with a starting point (immediately after Darwin) followed by an uninterrupted series of authors to the present. The author thus concludes: contemporary conceptions of human enhancement belong to a tradition of thinking that he characterises as an 'ideology of extreme progress'. Originally, this ideology was developed as a means of weakening the existing social order with its strong religious orientation. Today, with its astounding avoidance of social problems, this ideology functions as an antidote to social progressivism, and constitutes an 'opiate' for a significant fraction of today's middle class, especially scientists and engineers. This is indeed a radical assertion, which calls to mind the similar (but distinct) history of eugenics. Eugenics also appealed to a fraction of the middle class, in this case medical doctors and also teachers, who found in it a vision of society that concurred with their biologically oriented and meritocratic values.

In a totally different vein, Françoise Dupeyron-Lafay also suggests a pessimistic view of human enhancement. A specialist of 19th-century British literature, she highlights traces of the modern dream of human enhancement in Robert Louis Stevenson's novella, *The Strange Case of Dr Jekyll and Mr Hyde*. Today, this Victorian tale is mainly known through its numerous film adaptations, which unfortunately emphasise the dramatic physical and moral transformation of the main character, leaving aside the more complex psychological and metaphysical aspects of the novella. Dupeyron-Lafay insists on two aspects that are relevant to our reflection on human enhancement. First, she emphasises Jekyll's ambivalent but nonetheless confident belief in the powers of 'transcendental medicine', a scientific medicine capable of producing the miracle of freeing him from his limitations. This strange mixture of science and mysticism, Dupeyron-Lafay suggests, may also be at work in contemporary visions of the future of humankind. We cannot resist here, even if we go a bit further than the author, recalling the way Julian Huxley introduces his word 'transhumanism' in 1957: 'The human species can, if it wishes, transcend itself...in its entirety, as humanity. We need a name for this new belief. Perhaps *transhumanism* will serve: man remaining man, but transcending himself...' (full quotation in Bateman and Gayon, Chapter 1). The second aspect of this tale, which Dupeyron-Lafay mentions in her title, is Jekyll's conviction that the draught he takes will actually remove and isolate the negative aspects of himself, primarily the morally bad elements of his nature. The irony, of course, is that the drug produces a moral monster. Indeed, she points to the repressive social context of Calvinism that influenced Stevenson, and which probably accounts for Jekyll's obsession with purifying his person of his pleasure-seeking (thus evil) self. In sum, Stevenson's story can be read as a cautionary tale of human enhancement, with its combination of a desire for improvement and transcendence (rejuvenation, acquisition of new powers, surpassing our limitations, becoming 'good', and so on) and attendant fears about the process and its ultimate consequences (unreliability of our technical means, illusory scientific and technical powers, and the possibility of disaster).

In contrast with the pessimism of the two preceding chapters, the final two offer views that are the exact opposite. Both use fiction to develop their plea in favour of human enhancement, or at least to refuse its outright condemnation.

As Laure and Allouche did in Chapter 8, so Jean-Noël Missa addresses the problem of doping in sport. First he examines and challenges the 'anti-doping philosophy': in his view, all customary arguments are flawed. Missa does not defend indiscriminate and unregulated use of doping by athletes but, as a physician and historian of medicine himself, he argues in favour of an 'enhancement medicine'. In contrast with Jérôme Goffette (Chapter 2), he thinks that the boundary between therapy and medicine has gradually disappeared and that, especially in the case of professional sport, it is unrealistic, hypocritical and dangerous to ignore this. He gives several reasons why the practice of doping athletes should be recognised and medically managed. What appears to be his main argument is that an athlete is more than her genes and her own courage. Professional sport has become a highly competitive activity that, as such, relies on a combination of knowledge-based resources (professional training, hygiene, physiological knowledge, nutrition, environmental conditions, and so on) and in which drugs are one among many means to increase competitive excellence while alleviating unnecessary suffering. To support his views, Missa proposes an imaginary account of the Olympic Games held in Brussels in 2144. Suffice it here to say that Missa uses fiction to describe a state of affairs in which professional sport massively relies on the legalised use of doping, a situation in which this appears normal, morally acceptable and desirable for all agents – the public, sport institutions and the pharmaceutical companies sponsoring and coaching the athletes. At the end of the story, the narrator confesses that, although he has no doubt about the progress accomplished since the mad times during which doping was fiercely prohibited despite widespread use, he is not sure that he would like his son to become the kind of enhanced athlete that both father and son so enjoy admiring at the stadium.

Brian Stableford's final chapter confirms the interest of using fiction as a tool in making conjectures about the future of human enhancement. Initially trained as a biologist, Stableford has always been an admirer of John B.S. Haldane, a renowned biologist who occasionally engaged in speculative writing about the relationship between scientific progress – particularly in biology – and social change. Following in his footsteps, Stableford has, as a scientific writer, systematically explored the possible consequences of many forms of potential biological enhancements for individuals and societies. In his chapter, he points to the fact that he has always avoided apocalyptic scenarios: he has instead attempted 'to produce images of the future in which various kinds of biotechnologies are seen as entirely normal by the characters in the stories'. Here Stableford focuses on 'emortality' – a word that designates not immortality strictly speaking, but the scenario of an indefinite extension of the duration of life, in which death results only by accident. He tries to imagine its most probable effects on actual social attitudes. Those effects may occasionally be dramatic, but we need not assume that they will necessarily lead to tragedy, or that they are fundamentally undesirable changes. Perhaps his strongest example is his description of the effects of life extension on families and on the succession of generations: for example, the necessary changes in our rates of birth, in the ways we share the burdens and pleasure of having children and the possible consequences of these changes on our demographic and ecological environment. In our view, Stableford suggests that, if an enhancement is technically feasible and we find it desirable, then there are ways of re-organising society so as to cope with this new state of affairs.

So what is human enhancement all about?

As mentioned at the beginning of this introduction, the primary goal of this book is not to bear judgment on human enhancement, in the sense of arguing for or against it, or deciding whether or not it is ethically acceptable. Understanding it, rather than supporting or rejecting it, has been our central motivation in bringing together the authors who have contributed to this book. It is all the more interesting, therefore, to identify key points on which they agree and disagree. We will here insist on disagreements, because in our view these are the ones that structure the debate about human enhancement. Our intention is not to provide a systematic overview of all the issues that run through this book. Rather, we wish to highlight problems we did not initially anticipate or clearly identify, but that have since emerged in our discussions as issues needing more attention in future research on human enhancement.

Although virtually all authors have evoked the therapy-enhancement distinction and recognised its problematic status, most authors have not given this distinction a central role. However, two authors do give it greater importance but have divergent views about what practical consequences should be derived. Jérôme Goffette (Chapter 2), who admits that there exists a continuum of practices covering the two sides of the distinction, nonetheless claims that this distinction makes sense because it is extremely important in practice. Treatment and enhancement are indeed distinguishable from the viewpoint of their respective goals and values; each practice implies a distinctive sense of professional responsibility and, for that reason, should be entrusted to different professions – respectively, physicians and anthropotechnicians – each regulated by a specific set of norms and rules. Jean-Noël Missa (Chapter 11), on the other hand, also recognises the difficulty of distinguishing treatment and enhancement and, like Goffette, recommends a pragmatic attitude to this difficulty. But his solution is diametrically opposed to Goffette's, at least in the case of sport: medicine should embrace enhancement ('enhancement medicine'). This points to a real difficulty, which is not a theoretical problem: it is a practical problem that our societies will have to tackle, if we admit that enhancement is an acceptable practice. It might be relevant in dealing with this difficulty to point out that some, such as Missa, use a medically managed form of enhancement – such as doping – as their model, whereas others, such as Goffette, immediately encompass in their approach the possibility of enhancements that use non-medical technologies.

Our second reflection is about the social aspects of enhancement. Most authors support the idea that understanding human enhancement requires that we give due weight to its social dimensions. Winance *et al.* denounce the 'biotechnological illusion', that is, the belief that 'technological aids will *ipso facto* improve capacities'. Almost all authors emphasise the role of 'performance pressure' in enhancement practices. Bateman, Gayon and Menuz all insist that powerful social and cultural factors fuel and mould personal motivations to enhance oneself. Ter Meulen invites ethicists to include social concerns (especially solidarity) in their evaluation, and to avoid reaching too quickly the conclusion that all enhancement techniques and aspirations are necessarily selfish and narcissistic. Nevertheless, despite a quasi-consensus in favour of giving greater weight to social factors in current debates on enhancement, the last part of the book reveals two divergent views when it comes to imagining how enhancement technologies will affect the future of human societies. This divergence appears all the more clear when the debate moves to the political level. Whereas Coenen proposes thinking of current visions of enhancement as manifestations of an 'ideology of extreme progress', an ideology whose ultimate function is in reality to obliterate the notion of social progress, Stableford prefers to use fiction as a way of exploring the viability of a society of enhanced humans. This difference of perspective can be interpreted in terms of Karl Mannheim's distinction between ideology and utopia. Ideologies are systems of justification of a given social order. Utopias look forward to unexpected horizons. Utopias may be rooted in ideologies, but their distinct feature is to stimulate people to collectively act and think in radically new directions. The debate on human enhancement certainly contains strong ideological aspects, which should be explored more systematically than is usually done. But it also mobilises the power of utopia, when it explicitly addresses the social and political implications of radically new technologies.

All authors may not wish to endorse the conclusions that we have ventured to formulate, but we thank them warmly for having made them visible in the course of our collaboration.

Our critical inquiry initially included a section in which we hoped to learn something about human enhancement by examining the far older practices of animal improvement: we wished to compare them to present and prospective practices in humans. We hoped that the genetic improvement of animals, greatly reinforced today through recourse to zootechnics and biotechnology, might offer a paradigm that would enlighten us as to the future of human enhancement and the problems it might raise. The contributions to this investigation indicated that the idea of animal enhancement was a far more complex phenomenon than we had initially posited and merited an investigation in its own right: it emerged certainly as a possible prototype and but also as a countermodel of the approaches to human enhancement. We have therefore dedicated a smaller, separate volume to this issue – *Inquiring into Animal Enhancement: Model or Countermodel of Human Enhancement?* (Palgrave Macmillan, 2015) – so as to more fully investigate these contradictions.

Ultimately, this book and its companion volume on animal enhancement make a statement about the complexity of the concept of human enhancement and thus about the multiple perspectives required to examine it. Their overall spirit is to convincingly illustrate the interest of calling on specialists from various disciplinary environments to reflect on the nature, scope and issues at stake in the enhancement enterprise.

References

Agar N. (2004) *Liberal Eugenics: In Defence of Human Enhancement* (Oxford: Blackwell).
Buchanan A. (2011) *Beyond Humanity? The Ethics of Biomedical Enhancement* (Oxford: Oxford University Press).
Buchanan A. (2011) *Better than Human: The Promise and Perils of Enhancing Ourselves* (Oxford: Oxford University Press).
Bostrom N. and Savulescu J. (eds) (2009) *Human Enhancement* (Oxford: Oxford University Press).
Elliot C. (2003) *Better than Well: American Medicine Meets the American Dream* (New York: Norton).
Glover J. (1984) *What Sort of People Should There Be? Genetic Engineering, Brain Control, and Their Impact on Our Future World* (Middlesex, England: Penguin Books).
Harris J. (2007) *Enhancing Evolution* (Princeton and Oxford: Princeton University Press).
Hottois G., Missa J.-N. and Perbal L. (eds) (2015) *Encyclopédie du trans/posthumanisme: L'humain et ses préfixes* (Paris, Vrin).
Missa J.-N. and Perbal L. (eds) (2009) *'Enhancement'. Éthique et philosophie de la médecine d'amélioration* (Paris: Vrin).
Parens E. (ed.) (1998) *Enhancing Human Traits: Ethical and Social Implications* (Washington DC: Georgetown University Press).
President's Council on Bioethics (2003) *Beyond Therapy: Technology and the Pursuit of Happiness*, Foreword by Leon Kass, M.D., Chairman (Washington DC: US Government Office).
Rothman S. and Rothman D. (2003) *The Pursuit of Perfection: The Promise and Perils of Medical Enhancement* (New York: Random House).

Sandel M. (2007) *The Case against Perfection: Ethics in the Age of Genetic Engineering* (Cambridge, MA: Harvard University Press).
Savulescu J., ter Meulen R. and Kahane G. (eds) (2011) *Enhancing Human Capacities* (Chichester: Wiley-Blackwell).

Part I

Human Enhancement: What Do We Mean?

1
The Concept and Practices of Human Enhancement: What Is at Stake?

Simone Bateman and Jean Gayon

Human enhancement has generated a considerable amount of debate, with little concern as to *what* precisely is meant by this term and to what concrete practices it refers. Some authors, mostly philosophers, have offered formal definitions of the term; many others have often simply provided working definitions or based their reflections on an implicit understanding of the term (Menuz et al., 2013). We do not intend here to propose yet another definition, but to identify and characterize the contexts in which new uses of the term *enhancement* have emerged and to provide, on this basis, some points of reference as to what is at stake.

Our investigation, based on the collaboration of a sociologist and a philosopher and historian of science, led us to identify in the literature on human enhancement three distinct uses of the term, according to that aspect of humanness that is put forth as the aim of enhancement: improvement of human *capacities*, *self*-improvement and improvement of *human nature* (Bateman and Gayon, 2012). These three layers of meaning are not commonly distinguished in the literature, even though they are conceptually distinct; in fact, they tend to be amalgamated in conventional usage.

In trying to describe each one of these uses, we have tried to keep in mind the following questions: does the term human enhancement refer to forms of improvement that are novel and distinct from traditional *practices* for improving human beings? Does the term *human enhancement* imply changes in our *subjective views* as to what is meant by improvement? What *moral and political commitments* underlie contemporary controversies about human enhancement? Indeed, our analysis of the varying contexts in which present uses of the term have emerged has strengthened our initial intuition that the concept and practices of human enhancement are a specific and distinct phenomenon.

Improving human capacities

In 1993, LeRoy Walters, a philosopher and specialist in religious studies and ethics, gave a talk at the Hastings Center, a pioneering bioethics research institution in the United States, in which he raised questions concerning the moral acceptability of using 'genetic interventions' to enhance various human capabilities. He submitted 'four scenarios of genetic enhancement', an issue that he cautiously qualified as 'a taboo topic in current discussions of ethics and human genetics': genetic interventions might one day be used to improve the immune system, to decrease the need for sleep, to increase long-term memory, and to reduce the aggressive tendencies of humans while increasing their inclination to generosity.[1] Walters' thoughts on this issue, that he would later call 'enhancement genetic engineering' were part of a larger philosophical project that he was conducting on gene therapy (Walters and Palmers, 1997). His lecture provoked vivid controversies that went way beyond the issue of genetic enhancement. This led the Hastings Center to request and obtain, in 1995, funding from the National Endowment for the Humanities for a project entitled 'On the prospect of technologies aimed at the enhancement of human capacities'. The collective volume that resulted from the Hastings Center project – *Enhancing Human Traits*, edited by the philosopher Erik Parens and published in 1998 – was one of the first books to highlight this particular meaning of the term 'enhancement'.

The Hastings Center event thus appears as a significant milestone in the emergence of 'human enhancement' as a conventional term and an independent topic of ethical concern. There are two reasons for this. First, the event emphasizes the fact that the debate about practices later to be qualified as enhancements seems to have originated in the field of bioethics,[2] where there was already prior ethical debate on the social implications of biomedical practices derived from advances in the life sciences:

> The Institute [that is, the Hastings Center[3]] was founded in 1969 to fill the need for sustained, professional investigation of the social impact of the biological revolution. Remarkable advances were being made in organ transplantation, human experimentation, prenatal diagnosis of genetic disease, prolongation of life and control of human behaviour – and each advance posed specific problems requiring that scientific knowledge be matched with ethical insight.... The Institute's

approach was to bring together from many disciplines concerned professionals committed to meeting several times a year over a period of years, with 'homework' in between. These 75 Fellows...make up four groups, each concentrating on a specific area: (1) behaviour control, (2) population control, (3) genetic engineering and counselling and (4) death and dying. (Institute of Society, Ethics and the Life Sciences, 1973)

Indeed, the practices that we presently consider as forms of human enhancement are for the most part based on medical applications of biotechnology that lend themselves easily to a relatively stable functional classification, as exemplified by the table of contents of the book *Enhancing Human Capacities* (Savulescu et al., 2011): improvement of physical capabilities, cognitive abilities (perception, attention, memory and reasoning), mood, lifespan and quality of aging, and moral improvement. But what has made the enhancement of capacities a distinct issue is awareness that many therapeutic means can be used for non-therapeutic purposes (Rothman and Rothman, 2003; Goffette, 2006; see also Goffette, Chapter 2, and Menuz, Chapter 3, of this book). This is an old and familiar issue both in medical practice and in pharmacology, as illustrated by the case of amphetamines, originally developed in the context of research on molecules affecting blood pressure (Rasmussen, 2008). In the 1990s, the acceptability of non-therapeutic uses of medication became a regular issue in bioethical debates. What should one think, for example, about the prescription of growth hormone to children who are simply a little shorter than others? In 1996, 40 per cent of growth hormone prescriptions in the United States were for 'off-label' use (Sandel, 2007, p. 17). Should one use drugs, usually prescribed to slow the progression of Alzheimer's disease, to improve the memory of students or chess players (Frankford, 1998, p. 71)? These situations, and many others, have led to the development of a more systematic exploration of the distinction between the treatment of disease and the improvement of capacities – a constitutive element of the debate on human enhancement.

The Hastings Center event appears as a significant milestone for a second reason: it points to the fact that the debate about human enhancement seems to have originally emerged from bioethical discussions about the acceptability of *genetic engineering* (the third area of concern in the Hastings Center's original agenda). This practice, which evokes the perils of eugenics, was often addressed by philosophers in the field of applied reproductive ethics, such as Jonathan Glover (1984) or John Harris (1992). However, prior to 1990, the term *genetic enhancement* was not common in philosophical debates, even though one could find frequent reference to this term in the context of research on gene transfer in plant or animal microorganisms. But as the debate over human enhancement took hold, the meaning of the term evolved from an explicit reference to the controversial practice of genetic engineering to a notion encompassing a far broader range of practices, not limited exclusively to the biological sciences and biotechnologies.

Indeed, it is around the 1990s that the notion of technological convergence appeared, initially with reference to the digitalization of communication technology, and later extended to the other forms of technology. The term 'converging technologies' is now more frequently associated with the so-called 'NBIC revolution', that is the convergence of nanotechnology–biotechnology–computer science–cognitive sciences, as it is referred to in a report prepared by Mihail Roco and William Bainbridge for the US National Science Foundation in June 2002 (and later published in 2003):

> We are living through two tremendous patterns of scientific–technological change: an overlapping of a computer–communications revolution and a nanotechnology–biology–information revolution. Each alone would be powerful; combined, the two patterns guarantee that we will be in constant transition as one breakthrough or innovation follows another.
>
> Those who study, understand and invest in these patterns will live dramatically better than those who ignore them. Nations that focus their systems of learning, healthcare, economic growth and national security on these changes will have healthier, more knowledgeable people in more productive jobs creating greater wealth and prosperity and living in greater safety through more modern, more powerful intelligence and defense capabilities (Roco and Bainbridge, 2003).

In 2004, the European Commission also gathered a group of experts around the theme 'Foresighting the New Technology Wave'. Their report on this issue, published under the title *Converging Technologies – Shaping the Future of European Societies,* identified previous or ongoing research programmes involving converging technologies and defined the term as 'enabling technologies and knowledge systems that enable each other in the pursuit of a common goal.' (European Commission, 2004, p. 17).

Converging technologies, and the powerful research incentives behind them, have consequently expanded our overall conception of possible enhancement practices, thus releasing the concept of enhancement from its exclusive association with biomedical technologies, especially genetic, to which it had been attached until the 1990s. Indeed, it prepared the way for the dissemination of ideas and debate that go far beyond the bioethical debate.[4]

It is thus striking that even in as recent a book as *Beyond Humanity: The Ethics of Biomedical Enhancements* (2011), philosopher Allen Buchanan should choose *biomedical enhancements* as his principal concept; this seems to situate his own reflections within the bioethical tradition. The original emphasis on improvement of capacities appears clearly in his definition: '...interventions that directly improve human capabilities by the application of technologies to the human body or to human gametes or embryos.' (Buchanan, 2011b, p. 43). However, in a smaller book, *Better than Human: the Promise and Perils of Enhancing Ourselves* (2011), written and published later the same year, Buchanan distinguishes 'enhancement' as a general term from his concept of 'biomedical enhancement'. He defines enhancement itself as '...an intervention – a human action of any kind – that improves some capacity (or characteristic) that normal human beings ordinarily have, or more radically, that produces a new one.' (2011a, p. 5).

Even if Buchanan's two books provide an impressive analysis of the landscape of the human enhancement debate, his definitions do not fully capture the full extent of its practices. 'Human enhancement' as used today is indeed narrower than 'a human action *of any kind* – that improves some capacity...' (2011a, p. 5, our emphasis); however, with the emergence of converging technologies, the term embraces far more than biomedical practices. Nevertheless, Buchanan's terminology points to three important features of the modern debate on 'human enhancement': (1) the explicit aim to improve this or that capacity; (2) the idea that this requires direct and deliberate interventions on the human body; and (3) the idea that these interventions involve the use of technology.

Self-Improvement

In the same year that LeRoy Walters gave his lecture at the Hastings Center, psychiatrist Peter Kramer published a book called *Listening to Prozac* (1993). Kramer had been led to question the nature and effects of Prozac in depressive patients, after they began reporting that they felt not only relieved but 'better than well'. He coined the term 'cosmetic pharmacology', thus making an analogy with cosmetic surgery, to describe the use of psychopharmacology for personality modification. Kramer himself did not use the word *enhancement* in his book, but in 2003, philosopher and bioethicist Carl Elliot did, in a book entitled *Better than Well: American Medicine Meets the American Dream*. In this book, prefaced by Kramer, Elliot incorporates Kramer's views on psychopharmacology and extends them to a variety of 'enhancement technologies' aimed at improving one's appearance and sense of well-being (Prozac, Ritalin, Botox, Viagra, growth hormone, cosmetic surgery, sexual reassignment surgery, and so on).

Kramer and Elliot approach the human enhancement debate from a different perspective: their examples point to the notion of self-improvement, thus bringing to the debate the interpretation of enhancement practices by those who wish to make use of them. Self-improvement is not a structured doctrine but a subjective posture: the mental world of those who aspire to be enhanced is in most cases focused on the vision the individual wishes to give of him or herself, or as sociologist Erving Goffman (1959) would have said, on the presentation of self in everyday life. The technological devices are merely the means, the primary task at hand being work on one's self.

Carl Elliott's book, *Better than Well*, presents enhancement technologies in a cultural perspective: in his view, enhancement is first and foremost the continued pursuit of the 'American dream'. Citing 19th century historian and political philosopher Alexis de Tocqueville, he states that Americans, as opposed to Europeans, do not look to the past for guidance in making their choices but to their own judgment and to public opinion – an 'aspiration to self-sufficiency [that] easily slides into social conformity' (Elliot, 2003, p xix). Elliot's interpretation brings to mind sociologist Christopher Lasch's analysis (1979) of modern American culture, a 'culture of Narcissism', centred on the present and the satisfaction of immediate material needs.

However, Elliot also points out that people have an ambivalent attitude towards these technologies: despite cultural pressure towards self-fulfilment, using these technologies – even in situations where one might merely be seeking a solution to a real disability – entails the risk of experiencing a transformation of one's identity that may be perceived as far too radical. This is perhaps the most profound paradox of enhancement. On the surface, it is a variety of more or less invasive techniques that an unfamiliar observer might construe as a series of threats, transgressions, dangers or fantasies. But, according to Elliot, technical practices such as the use of sensory compensation by the blind and the deaf, or of an apparatus by those who have lost their capacity to speak, or even more radically, a change of sex for those who feel they were born with the wrong bodies, are far more than the mere improvement of existing capabilities: they are experienced as a quest for identity and authenticity, requiring difficult and perilous hermeneutic work on oneself (see also ter Meulen, Chapter 4 of this book).

The goal of self-improvement should not be mistaken with moral enhancement, currently the subject of much debate among philosophers, and already identified as a problem by LeRoy Walters in 1993, in the last of his four scenarios of human enhancement (see above). Some philosophers, such as Ingmar Persson and Julian Savulescu (2008), have explicitly endorsed the use of genetics and other biological means for 'morally enhancing human beings', for example by administering oxytocin to reduce aggression or increase the emotions associated with a sense of justice. Others, such as the philosopher John Harris who, as early as the 1990s, argued in favour of the moral legitimacy of genetic and more generally of human enhancement, does not approve of the idea of targeting 'specifically ethical capabilities.' In his view, only biotechnological interventions on cognitive abilities associated with the traditional means of moral improvement (socialization and education) are compatible with the freedom of individuals (2011). The debate over moral enhancement thus refers to the improvement of capacities, the first layer of meaning with respect to human enhancement; it is a normative debate as to whether or not this would be a desirable and acceptable practice. Self-improvement is a second layer of meaning that brings to the enhancement debate a subjective evaluation of its practices, viewed in terms of their impact on personal and social identity and a sense of well-being.

Although self-improvement may appear to be less radical as a goal than the attempt to increase one's physical, intellectual and moral capacities, it is sustained by an imaginary world that extends the possibilities of enhancement far beyond currently available practices. A striking example of this is the fact that the amputee sprint runner Oscar Pistorius developed the ambition to compete against able-bodied athletes, given the qualities of his carbon fibre prostheses. Speculations were long entertained as to whether they merely compensated his disability – the absence of legs below the knees – or whether they had qualities that made him a more efficient runner.[5] Riskier and more invasive experimental procedures for compensating physical disabilities are being developed that may generate similar expectations. Brain–machine interfaces, for example, presently make it possible for paraplegics to use their thought processes to control robotic arms, thus acquiring some limited mobility; but persons with no disabilities may wish to use BMI in the future to expand bodily sensations and extend their sphere of action (see Eskiizmirliler and Goffette, Chapter 7 of this book).

The wish to improve oneself also generates prospective visions of the self that emerge as cultural phenomena. Examples of these visions can be drawn from the artwork of artists such as ORLAN and Stelarc[6]: both have used surgical techniques, biotechnology and various other forms of modern technology to produce spectacular transformations of their bodies that challenge the norms defining the human body and explore the possibilities of human–machine hybridization. Whereas ORLAN's 'Carnal Art' focuses primarily on the religious, political, social and aesthetic norms that constrain the body's identity, in particular that of the woman's body, Stelarc explores the body's limits by amplifying its capacities through the use of prosthetics and robotics, and even through the use of human-machine interfaces that take over body movement, thus challenging the animated body as the seat of human autonomy.

Posthumanism as a recent trend in literary and cultural criticism has more generally addressed the way science and modern technology radically change our conceptions of embodiment and of subjectivity (Halberstam and Livingston, 1995; Hayles, 1999). Science fiction writers have also attempted to describe the personal experience of those with medically and technologically transformed bodies, as in the novel *The Dark Fields* by Alan Glynn (2001, later adapted to the screen as *Limitless*, 2011; see Zwart, 2014) – a trend whose roots can be traced back to Robert Louis Stevenson's *The Strange Case of Dr. Jekyll and Mr Hyde* (1886 – see Dupeyron-Lafay, Chapter 10 of this book), and possibly even to Mary Shelley's *Frankenstein* (1818). However, the specific characteristics of 'posthuman' science fiction began to emerge after the Second World War, when both writers and scientists expressed their disillusionment with humanism and its values dating back to the Enlightenment. Bernard Wolfe's dystopian science fiction novel *Limbo* (1952),[7] heavily inspired by the work of mathematician Norbert Wiener, is a major marker of this turn of events: the novel describes a society devastated by a nuclear third world war set off by computers to whom humans have delegated political decision-making, and in which the amputation of men's limbs becomes a form of 'moral enhancement' to curb aggressive tendencies. The imaginary world of self-improvement has since drawn freely from the new field called *cybernetics*, a term coined by Norbert Wiener in 1948 to refer to a general theory of information control and communication in animals and machines, based on formal analogies between the behaviour of living organisms and the operation of electronic and mechanical systems (Wiener, 1954).

However, not all contemporary science fiction writers feel that radically altered forms of human existence lead inevitably to dystopia. Some writers have organized their plots beyond the sole transformative experience of enhancement and tried to imagine what changes in social relations and in the physical and 'natural' environments will inevitably ensue from augmented human capacities – for example, how will longer life spans change the relationships between generations? Heightened human capacities and even radical morphological change do not necessarily preclude acceptable forms of human and social existence (see Stableford, Chapter 12 of this book).

The aspiration to surpass our 'natural' limitations, sustained by a rich imaginary world that draws its inspiration from a renewed vision of the role of science and technology in a world where the values of humanism seem to have failed, provides a strong anthropological foundation for this particular layer of meaning in the human enhancement debate.

Improvement of human nature

A growing interest in the impact of science and technology on humans and on social relations developed into more than the rich imaginary world of artists and writers. Philosophers, scientists and engineers began to think more concretely about the ways in which emerging technologies, notably the so-called NBIC revolution, could be implemented to improve the future of humanity, including human beings themselves. Two associations were created, about ten years apart, with the same intent: to promote the 'improvement of the human condition' as a valid goal for research and public policy.

The philosophy underlying the creation of a first association in 1991, The Extropy Institute, was elaborated in the late 1980s by the philosopher Max More and philosopher-lawyer Thomas W. Bell (More, 2013, p. 12). The term *extropy*, coined by Bell, was used not as a technical term opposed to entropy, but as a metaphor whose meaning was defined in the first version of the *Principles of Extropy* (published in 1990) as 'the extent of a living or organizational system's intelligence, functional order, vitality, and capacity and drive for improvement'. The seven principles of extropy – perpetual progress, self-transformation, practical optimism, intelligent technology, open society, self-direction and rational thinking – were advanced as an 'evolving framework of values and standards for continuously improving the human condition'. They were formulated 'to make sense of the confusing but potentially liberating and existentially enriching capabilities opening up to humanity', including that of 'morphological freedom' – the possibility of changing one's form. If the term extropy and the Extropy Institute (closed in 2006 – see Anissimov 2006) have since become historical relics in the enhancement debate, Max More claims in a recent article that these principles, and the Institute that promoted them, were nonetheless the first full expression of 'the core values and goals of transhumanism' (More, 2013, p. 5).

In 1998, another association, the World Transhumanist Association (WTA), was founded by philosophers David Pearce and Nick Bostrom. This brought to the debate a new term – transhumanism – to convey a more evolutionary perspective on enhancement. In a short introduction to transhumanism, Nick Bostrom presented it as 'a new way of thinking that challenges the premise that the human condition is and will remain essentially unalterable', and acknowledged the 'broadly technophiliac values' that underlaid this 'new way of thinking' (Bostrom, 2001 [1998], pp. 1, 6). In a later and extended version of this introduction, Bostrom, together with other colleagues,[8] refined the initial definition: transhumanism became the 'intellectual and cultural movement that affirms the possibility and desirability of fundamentally improving the human condition through applied reason, especially by developing and making widely available technologies to eliminate aging and to greatly enhance human intellectual, physical, and psychological capacities' (Bostrom, 2003, p. 4). The improvement of human capacities is here presented as subservient to a more ambitious goal: the evolution of humanity as such. Indeed, this later version claimed that 'the *human species* in its current form does not represent the end of our development but rather a comparatively early phase' (Bostrom, 2003 – our emphasis).

In 2008, the WTA changed its name to *Humanity+*. Following the closure of the Extropy Institute in 2006, *Humanity+* has since become the principal representative of the transhumanist movement. It presents itself as an international, non-profit organization 'dedicated to elevating the human condition':

> We aim to deeply influence a new generation of thinkers who dare to envision humanity's next steps. Our programs combine unique insights into the developments of emerging and speculative technologies that focus on the well-being of our species and the changes that we are and will be facing.[9]

This quotation reveals an intricate balance of differing modes of 'transhumanism': speculation about the evolution of the human species, and prospective thinking and planning about the use of emerging technologies.

Both Max More and Nick Bostrom recognize that transhumanism is not in itself a new term. Max More mentions anecdotal uses of *transumanare* by Dante Alighieri in the *Divine Comedy* (1312) and of the word *transhumanized* in T.S. Eliot's *The Cocktail Party* (1935) (2013, p. 8). But both More (2013) and Bostrom (2003, p. 41) rightly credit the renowned English biologist Julian Huxley for having coined this new term in the introductory chapter to his collection of essays, *New Bottles for New Wine*, published in 1957:

> The human species can, if it wishes, transcend itself – not just sporadically, an individual here in one way, an individual there in another way, but in its entirety, as humanity. We need a name for this new belief. Perhaps *transhumanism* will serve: man remaining man, but transcending himself, by realizing new possibilities of and for his human nature. (Huxley, 1957, p. 17)

However, if Huxley's language was in many ways very similar to present transhumanist discourse, he nonetheless saw transcendence as compatible with 'man remaining man'. Indeed, Huxley's analysis of the problem was quite different in that science and technology are subservient to social and even spiritual aims:

> We are already justified in the conviction that human life as we know it in history is a wretched makeshift, rooted in ignorance; and that it could be transcended by a *state of existence based on the illumination of knowledge and comprehension*, just as our modern control of physical nature based on science transcends the tentative fumblings of our ancestors, that were rooted in superstition and professional secrecy. To do this, we must study the possibilities of creating *a more favourable social environment*, as we have already done in large measure with our physical environment. (Huxley, 1957, p. 16 – our emphasis)

Huxley's transhumanism, or what he alternatively called 'evolutionary humanism' or 'the advancement of our species as a whole', referred to the creation of a social and technical environment that would favour a higher fulfilment of human beings and an improved quality of life. Indeed his ultimate goals – 'beauty', 'quality of people, not mere quantity', 'true understanding and enjoyment [as] ends in themselves', 'depth and wholeness of the inner life' – and the means he advocated – improvement of the physical environment, population control, education and 'techniques of spiritual development' – attest to this (Huxley, 1957, 16–17).[10] This vision of the future is quite different from that proposed by contemporary transhumanism, more focused on the constraints of the human biological heritage and on the technological means of transcending them, individually and as a species:

> Just as we use rational means to improve the human condition and the external world, we can also use such means to improve ourselves, the human organism. In doing so, we are not limited to traditional humanistic methods, such as education and cultural development. We can also use technological means that will eventually enable us to move beyond what some would think of as 'human'. (Bostrom, 2003, p. 4)[11]

These different versions of transhumanism all have a point in common: they bring to the enhancement debate a new layer of meaning – the aim of improving human nature. Rigorously speaking, human nature is not the term most frequently used by transhumanists; most often, they speak of 'the human condition', 'the human species' or 'humanity'. Indeed, changing and improving 'human nature' is a far more controversial claim, given that invoking the 'nature' of something classically involves the idea of a fixed essence which, by definition, cannot be altered. Nevertheless, we think that 'human nature' is the term most appropriate for describing this layer of meaning, because transhumanists challenge the inalterability of humanness. Moving humankind 'beyond what some would think of as "human"' (Bostrom, 2003, p. 4) announces a new chapter in the longstanding philosophical tradition of debate over human nature, in which the enduring relevance of the term is challenged.

Nor are transhumanists the only protagonists in this debate. As early as 1984, philosopher Jonathan Glover published a book, *What Sort of People Should There Be?*, about 'some questions to do with the future of mankind' (p. 13). In it, he considers how genetic engineering and 'brain control' (mood changes through drugs, brain stimulation and conditioning techniques) 'may change the central framework of human life'; he identifies changing human nature as a central issue, and argues in favour of a 'greater willingness to consider policies that would do this...' (p. 16). He nonetheless acknowledges that what one means by changing human nature is hard to pin down: 'The idea of "human nature" is a vague one, whose boundaries are not easy to draw. And, given our history, the idea that we must preserve all the characteristics that are natural to us is not obvious without argument.' (Glover, 1984, pp. 55–6). But Glover's main concern is that both 'changes in society and in human nature can be expected to involve changes in values' (p. 16). Technological interventions aimed at making these changes thus 'open up the possibility of producing a world of people with very different values' (p. 136).

In 1992, another moral philosopher, John Harris, expressed views about biotechnological enhancement that more closely echoed the emerging transhumanist movement:

> For the first time we can literally start to shape not only our own destiny in terms of what sort of world we wish to create and inhabit, but in terms of what we ourselves wish to be like. We can now, literally, change the nature of human beings. (Harris, 1992, p. 2)

In 2007, Harris went even further, stating that such practices were not only permissible, but should be considered a moral obligation:

> If, as we have suggested, not only are enhancements obviously good for us, but that good can be obtained with safety, then not only should people be entitled to access those goods for themselves and those for whom they care, but they also clearly have moral reasons, perhaps amounting to an obligation, to do so. (Harris, 2007, p. 35)

Nonetheless, Harris states that he has no affinity with the transhumanist agenda: 'a movement or quasi-religion which promotes, encourages, and indeed has as its objective the creation of a new species of "transhumans"'. For Harris, creating a new species is quite distinct from the moral duty to improve life, health and life expectancy (2007, pp. 38–9).

For other authors, however, the idea that human nature can and should be improved is rapidly categorized as incompatible with a religious vision of the world ('Playing God'), and even with a respectful attitude towards Nature. It is also associated with eugenic ideologies and policies, leading these authors to condemn human enhancement as a politically dangerous programme, threatening human rights. This political perspective permeates a number of national and international statements, reports or drafts, such as those concerning cloning and genetic modification (Juengst, 2009). These documents point out the risk of creating a biological gap between human populations who have access to enhancement technology and those who do not, with a concomitant risk of disqualification, oppression and bondage. The most extreme example of such a position is a paper by George Annas, Lori Andrews and Rosario Isasi, proposing that the United Nations vote a 'Convention on the preservation of the human species' that would outlaw all 'species-altering techniques':

> In fact, cloning and inheritable genetic alterations can be seen as crimes against humanity of a unique sort; they are techniques that can alter the essence of humanity itself (and thus threaten to change the foundation of human rights) by taking human evolution into our own hands, and directing it towards the development of a new species, sometimes called 'post-human'. (Annas et al., 2002, p. 153)

'Improving human nature' thus encompasses very different projects and visions of the future, often incompatible with one another. Indeed, in contrast with the preceding layers of meaning, the improvement of human nature generates a far more abstract philosophical, moral and political discussion that leads to endless polemics about the desirable goals and the appropriate means for the transformation of human individual and social existence. The peculiarity of this debate is that it takes as its object our very conception of 'what it means to be human', questioning its foundational value in our thinking about future worlds.

Conclusion

Our investigation of the semantic content of the term *human enhancement* and of the contexts in which these uses emerged suggests that the social, political and practical reality underlying the use of the term is far more complex than can be captured by the usual *pro* and *contra* pleas of the enhancement debate. It took several decades for the different layers of meaning to acquire shape and density, with the full landscape revealed in its breadth and complexity only at the turn of the Millennium. The three layers of meaning described above are rarely present in a pure form; they tend to overlap and intermesh to varying degrees, even in the writings of one author.

The notion that enhancement refers to the improvement of human capacities is the most visible layer of meaning, in that it is often associated with prototypic forms of enhancement (such as 'smart drugs').

These practices have so far arisen in the well-known area of biomedical practices, where technology assessment and the evaluation of results are routine procedures. Discussion as to what is ethically admissible or not in terms of 'human enhancement', understood as improvement of capacity, is therefore, in principle, closely scrutinized under a preexisting normative framework rooted in bioethics. How long bioethics will continue to provide a valid workable framework for evaluating practices based on a wider spectrum of technologies is however an open question. The longstanding tension in medicine between therapeutic and non-therapeutic uses of medication and medical procedures will undoubtedly play a key role in putting this framework to test.

The notion that enhancement is about self-improvement is not as much a debate as it is an open field for experimentation with concurring visions of human transformation: the aspiration to 'enhance one's self' motivates concrete practices and permeates the imaginary world of writers and artists who envision future societies in which our relationships to ourselves and to others will be profoundly transformed by pervasive technologies. These experiments affect our thinking about novel forms of embodiment and performance. They change our way of conceiving personal identity and our relationships to others. Our appraisal of these experiments in self-transformation will have a profound effect on the future of the human enhancement debate.

Finally, the notion that enhancement concerns the improvement of human nature and, more radically, the evolution of humankind towards a posthumanity, has become an increasingly important theme in the moral and political debate about the transformation of human society. The improvement of human capacities is extended to include the transformation of the 'human condition', and even that of the species. In other words, there is no essence of humanness that needs to be preserved. The debate at this level of meaning challenges the usual terms with which we engage in prospective thinking, and contests the limits habitually imposed to radical change of ourselves and of the physical and social environment.

Improving human *nature* is thus a radical scenario of the improvement of human *capacities* that extends its transformative vision to our commonly held views of human society. Whether or not radical enhancements will be perceived by individual members of that society as desirable, 'liveable' improvements of the *self* is likely to become a crucial arbitrating point in the debate over how far we wish to take the quest for human enhancement.

Notes

1. 'Ethics, Co-Creation, and Genetic Enhancement: A Response to Professor Gustafson's Paper', unpublished manuscript of a paper presented in Houston, Texas, on March 14, 1992, completed on August 13 of the same year. According to LeRoy Walters, 'the August 1992 essay that I sent you was the basis for my oral remarks at the Hastings Center.... Erik Parens cited passages from this essay in his introduction to "Enhancing Human Traits", which was not published until 1998' (Walters to Bateman and Gayon, e-mail dated April 4, 2012).
2. For a detailed account of the origins of this field, see: Fox and Swazey (2008), in particular Part 1.
3. The Hastings Center was originally founded as the Institute of Society, Ethics and the Life Sciences.
4. This link with converging technologies is explored in greater detail in Gayon and Bateman, 2014.
5. When Pistorius was finally allowed to compete with able-bodied athletes in the 2011 World Championships in Athletics and in the 2012 Summer Olympics, his performance in no way exceeded that of able-bodied athletes: he came in eighth and last in the second semi-final of the 400 metres race. (See Winance, Chapter 6 of this book).
6. http://www.orlan.eu/biography/; http://stelarc.org/?catID=20239. See also Steyn, 2005; Coulombe, 2009.
7. We thank Jérôme Goffette for pointing out the importance of this science fiction novel as a literary precursor in the human enhancement debate. See also Hayles, 1999.
8. Although this later version (2.1) of the Transhumanist FAQ was signed on the title page by Nick Bostrom, and can still be found online (http://www.transhumanism.org/resources/FAQv21.pdf), Bostrom was apparently the coordinator of a collectively redacted text, as indicated in the final section on 'Acknowledgements and Document History'. Some of the contributors were also members of the Extropy Institute. The most recent version (3.0) is on the Humanity+ site (as the WTA was renamed in 2008), and is presented as a collectively redacted text, with no primary author.
9. http://humanityplus.org/about/.
10. Huxley's penchant for prospective thinking developed within an intellectual circle that included the biologist J. B. S. Haldane, author of *Daedalus or Science and the Future* (1923), the writer H. G. Wells, best-known for his science fiction, and Huxley's own brother Aldous, author of *Brave New World*.
11. This passage remains identical in version 3.0 of the Transhumanist FAQ: http://humanityplus.org/philosophy/transhumanist-faq/#answer_19.

References

Anissimov M. (2006). "Extropy Institute Closes", Blog – *Accelerating Future*, May 5. http://www.acceleratingfuture.com/michael/blog/2006/05/extropy-institute-closes/, accessed March 24, 2014.

Annas G., Andrews L. and Isasi R. (2002) 'Protecting the endangered human: toward an international treaty prohibiting cloning and inheritable alterations', *American Journal of Law and Medicine*, 28, 151–78.

Bateman S. and Gayon J. (2012), 'L'amélioration humaine: trois usages, trois enjeux', in *Médecine/Sciences*, n° 10, 28 (octobre), 887-91. DOI: 10.1051/medsci/20122810019.

Bostrom N. (2001) [1998] *What is transhumanism?* Downloaded on April 22, 2012: http://www.nickbostrom.com/old/transhumanism.html.

—— (2003) The Transhumanist FAQ — A General Introduction (Version 2.1), World Transhumanist Association. Downloaded on April 27, 2012: http://www.transhumanism.org/resources/FAQv21.pdf.

Buchanan A. (2011a) *Better than Human: The Promise and Perils of Enhancing Ourselves* (Oxford: Oxford University Press).

Buchanan A. (2011b) *Beyond Humanity: The Ethics of Biomedical Enhancements* (Oxford: Oxford University Press).

Coulombe M. (2009) *Imaginer le posthumain. Sociologie de l'art et de l'archéologie et archéologie d'un vertige*, Préface de David Le Breton (Laval: Presses de l'Université Laval – Collection *Sociologie au Coin de la Rue*).

Elliot C. (2003) *Better than Well: American Medicine Meets the American Dream* (New York: W. W. Norton & Company).

European Commission (2004) *Converging Technologies – Shaping the Future of European Societies. A Report from the High Level Expert Group on 'Foresighting the New Technology Wave'*, Alfred Nordmann, rapporteur (Luxembourg: Office for Official Publications of the European Communities), 63p.

Frankford D.M. (1998) 'The treatment/enhancement distinction as an armament in the policy wars', in Parens, E. (ed.) *Enhancing Human Traits: Ethical and Social Implications* (Washington D.C.: Georgetown University Press), pp. 70–94.

Fox R.C. and Swazey J.P. (2008) *Observing Bioethics* (New York, NY: Oxford University Press).

Gayon J. and Bateman S. (2014), 'L'amélioration humaine (*human enhancement*)', in Carosella E.D., (ed.), *Nature et artifice: l'homme face à l'évolution de sa propre essence* (Paris: Hermann Editeurs), pp. 227–274.

Glover J. (1984) *What Sort of People Should There Be? Genetic Engineering, Brain Control, and Their Impact on Our Future World* (Middlesex, England: Penguin Books).

Goffette J. (2006) *Naissance de l'anthropotechnie. De la médecine au modelage de l'humain* (Paris: Librairie philosophique J. Vrin).

Goffman E. (1959) *The Presentation of Self in Everyday Life*. (Garden City, New York: Doubleday Anchor Books)

Glynn A. (2001) *The Dark Fields* (London: Faber and Faber).

Halberstam J. and Livingston I. (eds) (1995) *Posthuman Bodies* (Bloomington and Indianapolis: Indiana University Press).

Haldane J.B.S. (1924) *Daedalus or Science and the Future* (New York: E. P. Dutton and Co., Inc.).

Harris J. (1992) *Wonderwoman and Superwoman. The Ethics of Biotechnology* (Oxford: Oxford University Press).

Harris J. (2007) *Enhancing Evolution. The Ethical Case for Making Better People* (Princeton: Princeton University Press).

Harris J. (2011) 'Moral enhancement and freedom', *Bioethics*, 25(2), 102–11.

Hayles N.K. (1999) *How We Became Posthuman: Virtual Bodies in Cybernetics, Literature and Informatics* (Chicago: University of Chicago Press).

Huxley A. (1932) *Brave New World* (London: Chatto & Windus).

Huxley J. (1957) 'Transhumanism', in Huxley J. (ed.) *New Bottles for New Wine* (London: Chatto and Windus), pp. 13–17.

Institute of Society, Ethics and the Life Sciences (1973) *The Hastings Center Studies*, 1(1), Inside front cover.
Juengst E.T. (2009) 'What's taxonomy got to do with it? "Species integrity", human rights, and science policy'. In: Savulescu. J. and Bostrom, N, (eds) *Human Enhancement*. (Oxford: Oxford University Press), pp. 43–58.
Kramer P. (1993) *Listening to Prozac: A Psychiatrist Explores Antidepressant Drugs and the Remaking of the Self*. (New York: Viking Press).
Lasch C. (1979) *The Culture of Narcissism: American Life in an Age of Diminishing Expectations*. (New York: Norton & Company).
Menuz V., Hurlimann, T. and Godard, B. 2013. 'Is enhancement also a personal matter?', *Science and Engineering Ethics*, 19, 1, pp 161–77. DOI 10.1007/s11948-011-9294-y, published online 23 July 2011.
More M. (2013) 'The philosophy of transhumanism', in More M. and Vita-More N., *The Transhumanist Reader: Classical and Contemporary Essays on Science, Technology, and Philosophy of the Human Future* (Chichester: Wiley-Blackwell), pp. 1–17.
Parens E. (ed.) (1998) *Enhancing Human Traits: Ethical and Social Implications* (Washington D.C.: Georgetown University Press).
Persson I. and Savulescu J. (2008) 'The perils of cognitive enhancement and the urgent imperative to enhance the moral character of humanity', *Journal of Applied Philosophy*, 25(3), 162–77.
Rasmussen N. (2008) *On Speed: The Many Lives of Amphetamine* (New York: New York University Press).
Roco M. and Bainbridge W.S. (eds) (2002) *Converging Technologies for Improving Human Performance: Nanotechnology, Biotechnology, Information Technology and Cognitive Science* (National Science Foundation & Department of Commerce-Sponsored Report, Arlington, VA).
Rothman S. and Rothman D. (2003) *The Pursuit of Perfection. The Promise and Perils of Medical Enhancement* (New York: Pantheon Books).
Sandel M. (2007) *The Case against Perfection: Ethics in the Age of Genetic Engineering* (Cambridge, MA: Belknap Press of Harvard University Press).
Savulescu J., ter Meulen R. and Kahane G. (2011) *Enhancing Human Capacities* (Chichester: Wiley-Blackwell).
Shelley M. (1818) *Frankenstein or the Modern Prometheus* (London: Lackington, Allen & Co.).
Stevenson R.L. (1886) *The Strange Case of Dr. Jekyll and Mr Hyde* (London: Longmans, Green & Co.).
Steyn R. (2005) *Posthuman Body and Identity Modification in the Art of Stelarc and Orlan* (Dissertation submitted for Magister Technologiae: Fine Art, Tswhane University of Technology). Downloaded March 11, 2014: http://libserv5.tut.ac.za:7780/pls/eres/wpg_docload.download_file?p_filename=F1460946022/steyn.pdf.
Walters L. and Palmer, J.G. (1997) *The Ethics of Human Gene Therapy* (New York: Oxford University Press).
Wiener N. (1948) *Cybernetics, or Control and Communication in the Animal and the Machine* (Cambridge, Mass.: MIT Press).
Wiener N. (1954) [1950]. *The Human Use of Human Beings. Cybernetics and Society*, with an introduction by S.J. Heims (Boston, Mass.: Houghton and Mifflin).

Downloaded March 11, 2014: http://21stcenturywiener.org/wp-content/uploads/2013/11/The-Human-Use-of-Human-Beings-by-N.-Wiener.pdf.
Wolfe B. (1952) *Limbo* (New York: Random House).
Zwart H. (2014) 'Limitless as a neuro-pharmaceutical experiment and as a Daseinsanalyse: on the use of fiction in preparatory debates on cognitive enhancement', *Medicine, Health Care and Philosophy*, 17(1), 29–38.

2
Enhancement: Why We Should Distinguish Anthropotechnics from Medicine

Jérôme Goffette

Introduction: to distinguish or not to distinguish between enhancement and medicine?

'Enhancement', 'improvement', 'augmentation', 'alteration', 'moulding': such words, applied to humankind, cannot but bring to mind visions of science fiction scenarios. In fact, as the American President's Council of Bioethics noted in *Beyond Therapy* ten years ago, human enhancement is already a widespread phenomenon in today's societies. It is thus not surprising that studies, symposiums and projects on this topic are multiplying[1] (Missa and Perbal, 2009; Coenen, 2009). And what seems to emerge as a key issue in these debates can be summarised in one question: is it possible to distinguish between enhancement and medicine, between care that goes 'beyond therapeutic uses' and traditional medical care, between a desire to be 'better than well' (Elliott, 2003) and a desire limited to maintaining 'just health' (Daniels, 2008)? And if possible, should we act on that distinction?

One thing is clear: there is no consensus on the issue. Some authors contest any crude distinction between therapy and enhancement, supporting that enhancement is an extension of medical practice, or that there is a continuum between the two. Others claim that it is important and even necessary to maintain the distinction, for concrete or conceptual reasons.

But, behind this first question lies another line of inquiry, which is at least as important: an examination of the terms and concepts in play. It is interesting to note that, in France and continental Europe, scholars are at a loss when trying to choose an appropriate translation for the English term.

For example, Sfez (1993) talks about 'a utopia of perfect health', Sloterdijk (1999, 2000) and Hottois (Hottois and Missa, 2002) suggest the term anthropotechnics (or 'anthropotechniken' in German). Others simply use English terms like 'improvement' and, most often, 'enhancement'. Moreover they tend to reason in terms of a distinction between medicine and enhancement rather than in terms of the oft-referred to therapy-enhancement distinction; their focus is on differences in domains of professional activity rather than on the aims of specific practices (to cure or to enhance). In other words, when confronted with a new phenomenon, we hesitate over its name and meaning. Let us thus begin our academic endeavour, which aims to bring a sense of rigour and solid foundations to the discussion of enhancement and its associated concepts, by closely examining some of the main references on this topic today.

Beyond Therapy and *The Pursuit of Perfection*

The years 2002 and 2003 can be considered a key moment in the enhancement debate, with the publication of these four books:

- Fukuyama F. (2002): *Our Post-Human Future: Consequences of the Biotechnology Revolution*
- Elliott C. (2003): *Better than Well: American Medicine Meets the American Dream*
- President's Council on Bioethics (2003): *Beyond Therapy: Biotechnology and the Pursuit of Happiness. A Report of The President's Council on Bioethics*. Foreword by Leon Kass, M.D., Chairman.
- Rothman S. and Rothman D. (2003): *The Pursuit of Perfection: The Promise and Perils of Medical Enhancement*.

We will focus our inquiry on the last two, because they both focus on the 'therapy vs. enhancement' distinction and argue in favour of abandoning it. The two essays begin with lexical and conceptual considerations and advance several arguments to sustain their position. Let us examine their reflections in detail, by first looking at the principal arguments found in *Beyond Therapy* (President's Council on Bioethics, 2003), and then completing them with other arguments found only in *The Pursuit of Perfection* (Rothman and Rothman, 2003).

Argument 1 – 'Enhancement' is a non-specific and imprecise word

Part I, section 5 of *Beyond Therapy* is entitled 'The Limitations of the "Therapy *vs.* Enhancement" distinction'. Here is the first argument:

> Although the distinction between therapy and enhancement is a fitting beginning and useful shorthand for calling attention to the problem (and although we shall from time to time use it ourselves), it is finally inadequate to moral analysis. 'Enhancement' is, even as a term, highly problematic. In its most ordinary meaning, it is abstract and imprecise. Moreover, 'therapy' and 'enhancement' are overlapping categories: all successful therapies are enhancing, even if not all enhancements enhance by being therapeutic. Even if we take 'enhancement' to mean 'nontherapeutic enhancement,' the term is still ambiguous. (President's Council on Bioethics, 2003, p. 17).

Anyone who has listened to nurses and physicians know that the words 'enhancement' and 'improvement' are current, basic terms in a medical context. One 'enhances health', 'improves symptoms', 'heals or *enhances* the overall condition', and so on. Such expressions exist both in English and in many other languages. They express the concept of the transition from illness to health, from a pathological situation to an improved state and, at best, to recovery. 'Enhancement' is a widespread notion that includes positive therapeutic results. In fact, it seems impossible to establish a distinction between therapy and enhancement because a broad area of activities are therapeutic actions against illness and hope for an enhanced state of health.

Moreover, 'enhancement' has a *very* wide field of meanings. To get a rough indication, we looked at books containing 'enhancement' in their titles: they can be about improvement of semi-conductors, more-efficient management techniques, psychological advice about well-being, and biotechnical human empowerment. Only the last is really the topic of *Beyond Therapy*.

Thus, it seems that 'enhancement' cannot be clearly differentiated from therapy and is not therefore a specific word for the non-medical uses we have in mind.

Argument 2 – Defining the normal and the pathological, the normal and the enhanced, raises conceptual difficulties

The second argument against the therapy-enhancement distinction is stated in the following terms:

> There are difficulties owing to the fact that both 'enhancement' and 'therapy' are bound up with, and absolutely dependent on, the inherently complicated idea of health and the always-controversial idea of normality. The difference between healthy and sick, fit and unfit, are

experientially evident to most people, at least regarding themselves, and so are the differences between sickness and other troubles. When we are bothered by a cough and high fever, we suspect that we are sick, and we think of consulting a physician, not a clergyman. ... But there are notorious difficulties in trying to define 'healthy' and 'impaired', 'normal' and 'abnormal' (and hence, 'super-normal'), especially in the 'behavioural' or 'psychic' functions and activities. (President's Council on Bioethics, 2003, pp. 17–18)

In their essay, Sheila Rothman and David Rothman begin their criticism by pointing out the same problem: if it is impossible to express a clear separation between the normal and the pathological, then it is impossible to have a clear separation between curing and enhancing:

It turns out to be extraordinarily difficult, really impossible, to distinguish consistently between cure and enhancement, to compose a chart with interventions that aim to cure on one side of the page, and enhancements on the other. Cure itself is a highly ambiguous concept that requires precise definitions for such amorphous terms as health, disease, normal and abnormal, definitions that cannot be provided with any consistency or confidence. The most obvious example of how treacherous it is to invoke such terms comes from the field of psychiatry. (Rothman and Rothman, 2003, pp. xiii–xiv)

Firstly, when confronted with this argument, we must note that it is built on an analogical transfer: given that the distinction between the normal and the pathological cannot be consistently *established*, the same must be true of the distinction between the normal and the enhanced. However, even if philosophical controversies about the normal and the pathological are not closed, these terms are impossible to remove and replace. In medical practice, physicians constantly use them and, when considering medicine from a theoretical point of view, it is impossible to by-pass these words or find better ones. As a result, it is not because these words are difficult to define explicitly that they are not important, operative or even fundamental to medical thinking. We can use a metaphor: it is not because we cannot perfectly express what the colour blue is that it is not a consistent phenomenon. Besides, linguists know full well that common terms are both consistent and difficult to summarise or to define in simple terms.

Secondly, the *parallel* drawn between the two distinctions (pathological *vs.* normal and normal *vs.* enhanced) appears obvious at first glance but, when considered rigorously, is certainly false. The temptation is clearly to draw a single line tying the three notions together. Such a thought supposes that the meaning of 'normal' in the medical context is the same as in the 'enhancement' context. In the former context, 'normal' means the state of good health, a state the physician is aiming to restore. It is what we must call a limit-concept – the French *Code de Déontologie Médicale* clearly expresses this notion of a limit to a physician's acts (Ordre National des Médecins, 2012, art. 8, art. 40):

> [The physician....] *must limit* his or her prescriptions and actions to what is necessary to the quality, safety and efficacy of care.
>
> In the course of his or her investigations, interventions and prescriptions, the physician *must refrain* from submitting the patient to undue risks. (Our emphasis).

In the enhancement context, 'normal' does not mean a good physiological state but only one's condition at the beginning or at time zero, and we can note that this initial state need not be one of good health. For example, the same individual can be treated for an ulcer and consume psychostimulants in order to enhance his performance at work. Drawing a single line that links the three notions – pathological, normal and enhanced – also supposes that the two distinctions fit into the same scheme on both the semantic and the practical levels. In fact, even if the means may be similar, the ends are different and each end is conditioned by its own perspective. In conclusion, it is clear that the term 'normal' creates confusion, but the situation is neither a deadlock nor an absolute impossibility to conceptualise, and the situation may be open to clarification. Thirdly, if we take the example of psychiatry, the manifest difficulties encountered in attempts to characterise illnesses and relevant syndromes illustrate deep nosographic problems and we can observe some hesitation in drawing a distinction between pathological and normal phenomena. But it is equally true that these two concepts are constantly revised and that they fundamentally structure the physician's experience. The historical evolution concerning what is considered normal and pathological in psychiatry is not evidence in favour of the nonsense of these concepts, but evidence that they are fundamental concepts, even though their nosological content is not absolute and definitively fixed. Does a concept need to be absolute in order to be relevant? This would be absurd because, if such a condition is imposed, only mathematics could give us valid concepts.

Argument 3 – The World Health Organization definition of health implicitly includes both therapeutic and enhancing interventions

Beyond Therapy mentions the World Health Organization (WHO) definition of health:

> If one follows the famous World Health Organization definition of health as 'a state of complete, physical, mental and social well-being', almost any intervention aimed at enhancement may be seen as health-promoting, and hence 'therapeutic', if it serves to promote the enhanced individual's mental well-being by making him happier. (President's Council on Bioethics, 2003, p. 18)

Such a quotation, accurate though it may be, only supports an argument that appeals to the authority of an important international organisation. Even if this kind of argument must be taken into account as a sociological factor, it has no rational value in itself and must be examined in its content.

It is certainly true that if we follow the WHO definition, an important part of 'beyond therapy' enhancement can be seen as 'therapeutic'. Only enhancement without the purpose of attaining happiness would remain outside this definition. Thus, the question becomes: should we accept the WHO position? Its very generous definition of health, written in the post-war context, poses considerable problems. If complete well-being is the reference for health, then all of us would have to submit to medical treatment, because we are not in a state of 'complete' well-being. So, this definition seems too broad, inconsistent with its real aims and not operative. Moreover, by ignoring the role of the life context, it could sometimes appear absurd. The psychiatric model insists on the bio-psycho-social dimensions the physician should take into account for diagnosis and treatment. For example, being fired from one's job or in mourning for a dead friend involves sadness, so in these cases the individual being is not in a state of 'complete well-being', but such feelings are a – medically – normal reaction to the context. Conversely, feeling completely well and enthusiastic in a bad or average context can be the symptom of the euphoric stage of a bipolar disorder, a serious psychiatric condition.

Moreover the WHO definition is not clear. Is 'well-being' equivalent to happiness or simply the absence of pain or suffering? If it is equivalent to happiness, it would seem that a state of 'complete' well-being would be some kind of euphoria or a sense of plenitude. If it simply means an absence of pain, then it points to a more common disposition rather than to an extraordinary state of mind. Should a physician also be responsible for a patient's 'social' well-being or should this be left to politicians?

Therefore, even though official, the WHO definition must be treated with caution. Moreover, the feeling of happiness can have different meanings. On the one hand, it is a potential symptom or indication; on the other hand, it is the end of enhancement, or one of its ends.

Argument 4 – A conceptual distinction cannot be clearly established for properties distributed along a continuum

Here is the fourth argument:

> While in some cases – for instance, a chronic disease or a serious injury – it is fairly easy to point to a departure from the standard of health, other cases defy simple classification. Most human capacities fall along a continuum, or a 'normal distribution' curve, and individuals who find themselves near the lower end of the normal distribution may be considered disadvantaged and therefore unhealthy in comparison with others. But the average may equally regard themselves as disadvantaged with regard to the above average. If one is responding in both cases to perceived disadvantage, on what principle can we call helping someone at the lower end 'therapy' and helping someone who is merely average 'enhancement'? In which cases of traits distributed 'normally' (for example, height or IQ or cheerfulness) does the average also function as a norm, or is the norm itself appropriately subject to alteration? (President's Council on Bioethics, 2003, p. 18)

The argument of a continuum claims that, since there is no clear demarcation but rather continuity, there is no useful criterion for making any distinction. Against this argument of an imperceptible shift, one can argue that the existence of a continuum does not invalidate the reality of a polarity. The colour spectrum gives us an interesting metaphor: even if there is a continuum between red and violet, those two colours are clearly different. Rather, the reality appears polarised and one just establishes a classification in order to translate the continuity into degrees (as with discrete mathematical variables) that can be expressed by language.

If we refuse to admit the validity of categorisation in general, then we risk being incapable of thinking or taking any action at all. The syllogism can be continued: because there is always someone at a higher level for any capacity, and because each individual can legitimately demand to remedy a disadvantage, thus all individuals would have the right to remedy that disadvantage and to be equal to the more advantaged individual in all capacities. We hope such a conclusion appears as absurd to all. When we are speaking about health, we do not have in mind being as beautiful as a top model, as intelligent as a genius or as strong as Superman. We rather tend to have in mind a certain sense of absence of pain and the threat of being in pain or dying. The Aristotelian sense of 'meson' (mean position) or the common sense of the reasonable, however imprecise, is operative and meaningful. Even if there are intermediary situations, the polarity can be recognised and the distinction drawn.

Concerning the continuum argument, we can also repeat that curing and enhancing should probably be seen as two different dimensions, with two specific ends. Interpreting a bell curve as proof of identity bases the reflection on an epistemological mistake. For the physician, only the part of the curve representing persons with clinical risks or diseases corresponds to the field of medical action. For 'beyond therapy' enhancement, the curve can be seen merely as a distribution of inequalities, and potentially all the persons represented by the curve (even the best among them) could be improved.

Argument 5 – The distinction between the natural and the unnatural is anthropologically unfounded

In *Beyond Therapy*, the fifth principal argument against the distinction between enhancement and therapy is more developed than the preceding ones:

> Reliance on the therapy-versus-enhancement distinction has one advantage in theory that turns out also to be a further disadvantage in practice. The distinction rests on the assumption that there is a natural human 'whole' whose healthy functioning is the goal of therapeutic medicine. (President's Council on Bioethics, 2003, p. 19)

In the subsequent text, the reader is led to understand that seeing therapy as natural and enhancement as unnatural is a naive vision. On the one hand, therapy requires very artificial means and it fights against the natural course of events. On the other hand, the desire to improve oneself is a natural human wish and an anthropological aspiration. *Beyond Therapy* concludes that there is no legitimate general and global argument against 'beyond therapy' enhancement.

The *Pursuit of Perfection* puts this argument more clearly:

> For many critics, the core objection to the pursuit of perfection is that the results will be 'unnatural' and diminish our essential humanity.... Although an appeal to the natural has a long and venerable tradition – think of the elaborate formulations of natural law and natural rights – when put into the context of a specific human trait or condition, the yardstick falls short on several counts. First, there is no consensus in biological terms as to what defines the critical elements of our humanity, what constitutes 'the natural' that should be respected, or the 'unnatural' that should be avoided. By what grounds is it 'natural' to extract a kidney or a heart from a person who has just died and transplant it into a living being?... Why would it be unnatural to extend longevity to 140 years? (Rothman and Rothman, 2003, p. xiv)

This final argument is based on an almost unquestionable reality: the human world has always been a world of artificial affirmation. Anthropologists have often highlighted that the first gesture of humanness is to affirm its particularity, for example by performing an act that transforms natural bodies into humanised bodies (style of gestures, emblems, garments, make-up, body painting, hairstyle, and so on). Both medicine and enhancement belong to that accomplishment of humanness. We must agree that the naturalist argument appears today as outdated and simplistic. In fact, the natural *versus* artificial argument disguises a reference to the sacredness of the state of nature or a religious obligation to preserve the human shape derived from divinity.

Consequences of our critical examination

Given these arguments against the distinction, is it possible to find an argument in favour of the distinction? Why try to find such an argument? The first reason is the ambiguous attitude of *Beyond Therapy*. It explains that the distinction between what is therapy and what is 'beyond therapy' is 'inadequate' (President's Council on Bioethics, 2003, p. 17) and 'problematic' (President's Council on Bioethics, 2003, p. 18) and that it must be abandoned. But the report constantly refers to the distinction and even inscribes it in its own title: *Beyond Therapy*. The contradiction is disconcerting. The second reason partly explains why the report continues to use it. When it elaborates on this kind of activity, it adopts a methodological approach to the subject which is radically different from the approach physicians use to describe therapy. Therapy is developed in specialities related to injured organs (cardiology, dermatology, digestive medicine, ophthalmology, and so on) and according to the kind of pathological anomalies involved (allergy, cancer and infectious diseases, for example). This organisation is both official (with academic medical degrees) and effective (see the departments and services of a hospital). When one uses the word 'therapy', this structure, for reasons of history and the need for efficacy, is always present, even if it remains in the background. However, for its approach on biotechnological enhancement, the Council writes: 'Structure of the Inquiry: The Primacy of Human Aspirations' (title of part 1, VII) and it explains this choice in the following terms:

> We could begin from the novel *techniques*.... We could begin with the new *powers* or *capacities* these techniques provide.... We could begin with those *aspects of human life* that might be affected.... We could begin with the *desires and goals* that either drive our pursuit of these techniques or that will enlist the available powers they make possible once they are available: desires for longer life, finer looks, stronger bodies, sharper minds, better performance, and happier souls – in short, with our specific aspirations to improve our lot, our activities, or the hand that nature dealt to us or to our children.
>
> In keeping with our goal of 'a richer bioethics' – one that seeks to do justice to the full human meaning of biotechnological advance – we will here proceed in the last of these ways. (President's Council on Bioethics, 2003, p. 24)

If the distinction were really false, and the unity true, why did the Council make this choice? Highlighting this point is important, because this choice concerns their method of investigation, translates the particularity of the topic, and finally shows once again the contradiction between the refusal of the distinction and the way the Council has chosen to approach enhancement as a situation distinct from therapy.

In conclusion to our critical examination, only the first and a part of the second argument appear sound:

- 'enhancement' is a non-specific and imprecise word;
- leaving aside the medical concepts of the normal and the pathological, the concepts defining 'enhancement' should not be derived from a distinction between the normal *versus* the enhanced, as they are too imprecise, non-specific and ambiguous.

Medicine and anthropotechnics

We now have to answer the following question: does the word 'enhancement' cover a homogeneous field of activities? What practices do we have in mind when we talk about enhancement, improvement, pursuit of happiness, aspirations to be 'better than well', biotechnologies that are 'beyond therapy', smart drugs, and so on? Is there a consistency throughout these practices, for consistency is a necessary condition in identifying a concept.

Field and denomination: human enhancement and anthropotechnics

In a book published in 2006, based on a specific methodology, we attempted to map out these extra-medical practices, highlighting those seen as problematic and atypical (Goffette, 2006). The resulting inventory can be divided into eight kinds of practice:

- performance doping (legal or illegal);
- uses of psychostimulants (apart from medical indications);
- aesthetic transformations (apart from reconstructive plastic surgery);
- reproductive control (except in cases of reproductive dysfunction);
- mood modification (apart from medical indications);
- sex reassignment;
- quests for youth or immortality; and
- fabrication of a human being.

Conceptual analysis has shown that they involve:

- a unity in terms of matter or object: the human being;
- a specific end or finality: modifying the human being in order to improve performance (for example doping), to give more freedom (for example contraception) or to voluntarily transform self-identity (for example cosmetic surgery); and
- a specific problem concerning the regulations, rules, professional norms and laws supervising or controlling them.

All of these activities or projects follow the logic of improvement and modification, through mastery of the body. They provide services that go beyond normal medical health care, or, more correctly, they take another direction.

If we accept the definition of a discipline or profession as the combination of an underlying anthropological motivation, a set of activities expressing this fundamental motivation, and a normative framework (institutions, definitions and regulations), one must recognise that we are witnessing the birth of a new discipline. At least this is our hypothesis.

This new area, which is not biology (because its purpose is not knowledge about life), not biotechnology (because biotechnology is only a means towards various ends and applications) and not medicine (because it does not fight against illness), is very specific, with a particular finality: to give us the power to change ourselves, improve facets of our being or choose certain traits.

So we are led to the following hypothesis: a new kind of activity is being born, that is to say, the art of modelling, shaping the human being, the art of changing and improving one's being by bodily modification for non-medical purposes (Goffette, 2006, p. 169). Here, the denomination and the concept of 'anthropotechnics' can be appropriately applied. We have to note that, before our own use of this term, an adjectival form of the term had been proposed by the German philosopher Peter Sloterdijk (1999, 2000) and by the Belgian philosopher and ethicist Gilbert Hottois (Hottois and Missa, 2002).

The choice of an appropriate term for this new activity is not simply a superficial question of labelling. In the context of a cross-cultural discussion, talking about 'human enhancement' is normal but is nonetheless the source of numerous problems. In particular, we need to consider two issues:

- In the North American or British context, talking about 'enhancement' or 'enhancing technologies' appears easy but generates confusion. These words are common in English and are known to everyone. However, as highlighted by the President's Council of Bioethics and by Sheila and David Rothman, 'enhancement' is too vague a word, making it difficult to comprehend the issue. Talking about 'anthropotechnics' and 'anthropotechnical techniques' would clarify the situation, by putting aside the common words whose meanings would distract us from the important issues, and by avoiding the confusion created by the medical uses of the word 'enhancement' (for example 'enhancing care', 'enhancing a prognosis', and so on).
- In non-English speaking contexts, talking about 'enhancement' introduces several biases. Firstly, it provides an appearance of modernity – with the American language as the emblem of a technological future. Consequently, it induces a value judgment. For academic work, the effect of fashion and of its ensuing value judgments is more likely to be a source of trouble than clarity. Secondly, using the English word in a French- or German-speaking context transforms the meaning. One

moves from an everyday word to an erudite word, from a polysemic term to an apparently monosemic term, from a common notion to an attempt to establish a clear concept, and so on. So, it seems preferable to find another word for use in non-English speaking countries, even if only as an indicative label; 'anthropotechnics', separately reinvented by three authors, appears to be a good candidate ('anthropotechnie' in French, 'Anthropotechniken' in German, 'antropotecnica' in Spanish and Italian, and so on).

Fundamental concepts: the ordinary and the modified

Following these terminological considerations, we would like to develop the conceptualisation of anthropotechnics itself. As in medicine, a pair of polar concepts characterises anthropotechnics. But, whereas medicine is structured around the pathological and the normal, anthropotechnics apparently opposes the normal and the improved or the enhanced – as we saw in *Beyond Therapy*.

Yet, there is a problem concerning these concepts. The psychiatrist Édouard Zarifian distinguishes, when talking about psychotropic drugs, on the one hand the medical resolution of pathological problems and, on the other hand, the treatment of 'existential' problems (Zarifian, 1994, p. 196). This latter expression is used to explain the consumption of drugs by individuals who have no apparent psychiatric problems, given their bio-psycho-sociological situation.

First of all, we are faced with a necessary preliminary task: trying to define as far as possible the medical concepts of 'normal' and 'pathological'. This is a difficult and complex exercise. Nevertheless, it seems to us (for reasons that we are unable to develop here) that certain types of definitions are inadequate or inappropriate, such as the World Health Organization's notion of 'complete well-being', the reference to biostatistical norms, Canguilhem's concept of vital normativity, or the identification of the normal state with sociological normality. On the contrary, we wish to highlight, as Leriche (1966) has already done, the suffering and/or the pain linked to pathological states. But we have to separate pathological suffering from the existential suffering mentioned by Zarifian. The latter, such as the 'broken heart' after ending a relationship, the suffering from being 'ugly', or the sadness of bereavement, are medically normal, because they are a normal affective response to the life context. As a result, we have defined medicine as 'the activity whose purpose is to know, to prevent, to cure the pathological' and the pathological as 'the possible threat or the effective presence of an unexpected, inappropriate, disturbing and/or painful expression of the functions' (Goffette, 2006, p. 109). These definitions are neither revolutionary nor sophisticated, but they specify certain elements, underlying a common idea of these two concepts widely shared by patients and physicians alike, and that usually remain tacit or are considered self-evident.

With respect to this definition, we wish to emphasise the fact that certain anthropotechnical actions do not even treat medical or existential suffering, but aim at satisfying desires, or fulfilling professional requests. Moreover, anthropotechnical acts may sometimes generate suffering and may carry a risk without any benefit from the medical point of view (for example certain risks due to doping). Anthropotechnics is sometimes the opposite of medicine.[2]

We now need to elaborate the key anthropotechnical concepts, taking into consideration these key points in comparison with the medical conceptual framework. At first glance, anthropotechnics may appear to be constructed around a distinction between the normal and the improved. However, after careful consideration, this vision in terms of a pathological-normal-improved continuum appears to be inappropriate, as we have already seen:

- First, anthropotechnics is not a medical improvement (or enhancement), but a medical risk. We have in mind the medical balance between health benefits or benefits in the fight against disease (in these cases, it is zero because this is not the purpose, so any benefit is merely coincidental) and risks (since there is a modification of the body, we can observe various side-effects and associated risks). For example, there is a risk of dependence on psychostimulants in the case of doping, or anaesthetic risk during cosmetic surgery.
- Second, medicine uses the concept of medical improvement (or enhancement), when, for example, one speaks about improvement of the symptoms, in order to signal alleviation of suffering or disappearance of a physiological dysfunction. 'Improvement' is not specific to anthropotechnics.
- Third, 'improved' and 'enhanced' are words with positive connotations: we need a descriptive word without axiological implications or tacit values. For all these reasons, it seems preferable, in our view, to use the *ordinary* and the *modified* as anthropotechnical concepts. On the one hand, the baseline is not the normal in the medical meaning of this term, but rather the *ordinary* condition of a person, not necessarily healthy. On the other hand, speaking of improvement or enhancement is in fact inexact, because certain modifications pursue other aims: freedom, affirmation of identity, remaining competitive at school or at work, and so on. For example, a sociologist mentioned the case of a man who wished to change the shape of his nose in

order to have one different from his father's, even if the new nose was less attractive: he was seeking a symbolic break with a father who had abandoned him (Châtelet, 1993, p. 139). The concept of the *modified* is broader than improvement of performance and it can allow us to focus more clearly on the motivations behind the acts. Moreover, it is purely descriptive and it allows us to separate the description of the anthropotechnical phenomenon from the ethical position about its various forms.

Such a conceptual framework can have real consequences in concrete situations. One of them is the status of the modified. For medicine, the normal is a limit-concept, a concept that defines a limit beyond which medical practitioners should not act. For anthropotechnics, the modified does not set a limit but instead opens an infinite horizon, as wide and limitless as human desires. This characteristic can explain why, in order to understand what anthropotechnics means, the better methodological approach is *via* people's goals or aspirations. Finally, it explains why *Beyond Therapy* adopted such a specific approach and why, in opposition to standard medical books, it insists on the dimension of the dream. The dream is a constant feature of part 1, VI:

> Until now these dreams have been pure fantasies, and those who pursued them came crashing down in disaster. But the stupendous successes over the past century in all areas of technology and especially in medicine, have revived the ancient dream of human perfection. (President's Council on Bioethics, 2003, p. 21)

Several paragraphs further on, we can read a passage strangely close to our point of view:

> These dreams have at bottom nothing to do with medicine, other than the fact that it is doctors who will wield the tools that may get them realised. They are, therefore, only accidentally dreams 'beyond therapy'. They are dreams, in principle and in the limit, of human perfection. (President's Council on Bioethics, 2003, p. 22)

Even if the topic of perfection appears to us much too narrow, the idea of technical and anthropological reverie sounds right. While medicine must necessarily confront painful realities, anthropotechnics is rooted in dreams. The modified is not a limit-concept, but a horizon-concept, pushed indefinitely further.

Some pieces of evidence

When evoking the birth of a new domain, a new discipline, it is useful to provide both conceptual and concrete arguments in favour of this initiative. Thus, we will present some pieces of evidence in favour of anthropotechnics. Among the most interesting are the books aimed at the general public concerning the use of psychostimulants or 'smart drugs'. By 2006, no fewer than six books on this theme, including some bestsellers (*300 médicaments*, 1988; Pelton and Clarke-Pelton, 1989; Dean, Morgenthaler and Fowkes, 1993; Souccar, 1997; Sahelian, 2000; Lidsky and Schneider, 2001), had been published, with others following.

In France, two books promoting doping were particularly successful. The first was an anonymous publication entitled *300 drugs for physically and intellectually surpassing oneself*, published as early as 1988. The first chapter 'The right to use stimulants' is very direct:

> This work is not a book of medicine for ill persons. It is an informative tool intended for all men and women who feel the legitimate desire to improve their physical and intellectual performance, fight the fatigue and stress inherent in modern life. Taking drugs in order to resist, to express oneself, to win in the competition for life, has become a need felt by the majority. (*300 médicaments*, 1988, p. 9)

The difference between medicine and doping is clearly affirmed, against the background of a perceived underlying, generalised social demand. Incidentally, the authors judge this desire for improvement as 'legitimate'.

This 'legitimacy' can be better understood by looking at another quotation:

> Of course it is indispensable to remain in good health, well-balanced and to feel good about oneself. Keeping us in good health, curing us if we are ill: that's the role of predictive and healing medicine. But a healthy individual may desire or may need sometimes to do more, to surpass himself or herself, to outdo his or her physical and intellectual abilities. ... The conditions of modern life, frantic competition of candidates to obtain diplomas, jobs, success, better professional positions or affective gratifications make it indispensable to resort to invigorating and stimulating products. (*300 médicaments*, 1988, pp. 15–6)

Such a book emphasises a certain vision of modern life as a 'struggle for life'. In fact, the foundations of the legitimacy of doping are a strong version of the winner's ideology. For example, sports competition appears explicitly as a model for understanding modern society (*300 médicaments*, 1988, p. 16). In the book, the struggle model is exported to all other areas of life (*300 médicaments*, 1988, pp. 17–8)[3]: the school system, professional life, leisure and even personal relations ('affective gratification' is discussed in a chapter about aphrodisiacs and sexual stimulants). In 1988, the radical position of this book offended the medical world (Ordre National des Médecins, 1988a, b) and, three months later, the publishing company withdrew the book from the market. But by then, 150,000 copies had been sold, making it a great success in France.

The second French book of this kind was published in 1997, with similar success but without any reaction from the medical community (Souccar, 1997). Against a background of capsules, tablets and other pills, the following text appears on the cover (my translation):

Thierry Souccar
The Guidebook of New Stimulants
More effectiveness / More intelligence / More concentration
More energy / More optimism / More sexual tonus / More creativity

This book cover illustrates a central part of the anthropotechnical paradigm: these activities do not merely aim to fight against the 'less' – like medicine – but in favour of the 'more': 'more effectiveness', 'more intelligence', and so on. Anthropotechnics is not restricted by the limit of the normal, but opens itself up to the limitless horizons of human desires, social pressure and maybe even 'super-humanity' (see also Menuz, Chapter 3 in this book).

Of course, the English-speaking public has no shortage of books on these same subjects. Here is a similar book: *Smart Drugs II*, with the following texts on its front and back covers (Dean, Morgenthaler and Fowkes, 1993):

Front cover:

Smart Drugs II – The Next Generation
New Drugs and Nutrients to Improve and Increase Your Intelligence

Back cover:

Enhance your clarity of mind and sensory awareness
Increase your sexual enjoyment
Enhance school and job performance
Learn the latest treatments for Alzheimer's disease

Increase your energy and alertness
Increase your IQ by 10 points or more
Improve your problem-solving abilities
Improve your memory by as much as 40%
Slow down ageing

The argumentation used to promote the book is similar. Indeed, all these books emphasise the difference between their approach and that of medicine in their introduction and repeat it in their exploration of the pharmacopoeia: 'The benefits for "normal" people are measurable' (Pelton and Clarke-Pelton, 1989, p. 14); 'This book is not a medical book meant for ill persons' (*300 Médicaments*, 1988, p. 9); 'Piracetam Update. This unique substance is probably the most popular smart drug for normal, healthy people' (Dean, Morgenthaler and Fowkes, 1993, p. 103); or 'Men and women need to be alert and focused at work in order to maximise performance and reap the rewards of accomplishment and success. The appropriate use of mind boosters can help in this regard.' (Sahelian, 2000, p. 193).

In the USA we can find specialised books not only on psychostimulants, but also on other fields of anthropotechnics, such as *Performance-Enhancing Substances in Sport and Exercise* (Bahrke and Yesalis, 2002) for physical doping and body building, or books on cosmetic surgery aimed at the general public and indicating that the demand must be 'sane' and the patient healthy: 'We do not deal with ill but with healthy persons and that is our fundamental difference' (Hagège 1993, pp. 13–14), and so on. We also can find prevention books such as *Amphetamines and Other Stimulants* by Lawrence Clayton (1993), and so on.

So, for promoters and users themselves, the difference between medicine and what they are talking about appears clear and is acknowledged as providing a useful framework. Thus it is reasonable to believe that our analysis and associated conceptual work are not simply the abstract vision of a philosopher in his ivory tower, but a more clearly conceptualised expression of a widespread distinction made by practitioners and promoters of the relevant practices.

An operative distinction

The deontological problem

The epistemological distinction between medicine and anthropotechnics involves a deontological distinction. If we examine the issue carefully, we can see that there is an obvious conflict of values between the two areas. Medicine stands ideally for a pain- and disease-free life. Anthropotechnical acts may sometimes generate pain and represent a medical risk, but they promise no medical benefits. Moreover, as we have already pointed out, anthropotechnics sometimes operates in a direction quite opposite to that of medicine. Indeed anthropotechnics sometimes runs counter to the Hippocratic invocation 'Primum non-nocere', or to the more general principles of non-maleficence and medical beneficence (Beauchamp and Childress, 2008).

Such a situation poses concrete problems. For example, in France, the jurisprudence concerning aesthetic surgery is unclear and sometimes incoherent, and the law allowing the termination of pregnancy involves *ad hoc* articles in the medical code of ethics (such as a conscience clause that goes against the traditional medical imperative to perform medical acts). To turn to a different field of medical deontology, anti-doping laws are being enforced all over the world and are constantly under discussion, while the problem of doping at school and in other educational contexts – once almost unheard of – now seems to be on the rise.

Finally, the distinction we are drawing between medicine and anthropotechnics implies, we believe, a need to separate what appears to be two distinct professions. Furthermore, this would imply two different sets of professional codes of ethics and laws. Even if the same practitioner works in the two fields, it is important to clearly differentiate the two roles and their accompanying framework of responsibilities. Besides establishing clearer deontological standards and rules for physicians, we would have to devise a professional code of ethics for the practice of anthropotechnics, based on distinct principles, possibly including new forms of the principles of autonomy, respect of human dignity, the meaning of humanity, and so on.

The sequence of consultation

Another concrete example of the tension between medicine and anthropotechnics is offered by the different scenarios for a consultation. Traditionally, in medicine, a consultation follows this sequence:

- taking of the patient's medical history;
- auscultation;
- diagnosis;
- treatment and prognosis.

It is clear that this pattern is completely unsuited to anthropotechnics, because there is no illness to diagnose. The medical practice of establishing the patient's history aims to identify the disease through the symptoms. In anthropotechnics, there is no disease and no symptoms are presented, but rather a demand, a desire, or a wish for 'more'. In anthropotechnics, the word 'diagnosis' makes no sense. Even though it might be possible to talk about treatment in this context, anthropotechnics does not cure an illness with appropriate treatment, but simply suggests a certain choice of realisations, indicating risks and expected results. There is, of course, always the possibility of choosing to do nothing.

Here is a comparable sequence for a consultation in the context of anthropotechnics:

- expression of the client's request;
- auscultation in order to assess the risks and the contraindications;
- presentation of the various means available to satisfy the request, and explanation of the physical and psychological implications and the pathogenic risks relevant to each possibility;
- discussion concerning the 'real' request (making explicit a possible hidden reason for the request and consequently a reformulation of that request) and information about the implications of the various modifications;
- time for reflection;
- final choice by the client;
- technical intervention.

We should note that in this case, the 'patient' is not even a patient, in the etymological sense of this word, but rather a client. Furthermore, the physician is not considered a physician, but rather a service provider who must follow specific rules due to the risks generated by interventions that modify the bodies of human beings. Thus, the professional relationship engaged in anthropotechnics differs from the medical relationship between doctor and patient. The distribution and nature of ethical responsibilities is also very different, with practitioners being under no obligation to perform the services they offer, for example.

Conclusion

In conclusion, we want to emphasise the fact that the therapy/enhancement debate is far from closed and the discussion appears to be just at its beginning. Based both on our conceptual analysis of that distinction and on a study of empirical examples, we argue in favour of a sharp distinction between medicine and anthropotechnics.

Of course, it is quite obvious that ambiguous situations, combining medical and anthropotechnical dimensions, exist. Nonetheless, we believe that the distinction can be useful in providing a clarification of concrete practices, with their respective ethical approaches, thus helping to prepare for future developments in both fields.

Notes

1. Here are some examples of European meetings (less well known than their American counterparts): Brussels, May 9–10, 2008: Symposium on 'Enhancement – Aspects éthiques et philosophiques de la médecine d'amélioration' [Enhancement – Ethical and philosophical aspects of improvement medicine]; Paris, 2009–2010: Five workshops on 'Human Enhancement: an Interdisciplinary Inquiry', University Paris 1/University Paris 5/University Paris 7, at the origin of this book; Bristol 2011: beginning of the EPOCH project – *Ethics in Public Policy Making: The Case of Human Enhancement*, funded under the European Union's 7th framework program (see: www.epochproject.com).
2. This question is also discussed by Laure and Allouche in Chapter 8 of this book.
3. See also Laure and Allouche, Chapter 8 in this book.

References

(1998) *300 médicaments pour se surpasser physiquement et intellectuellement* (Paris: Balland).
Bahrke M.S. and Yesalis C.E. (eds) (2002) *Performance-Enhancing Substances in Sport and Exercise* (Champaign, Ill.: Human Kinetics).
Beauchamp T. and Childress J. (2008) *Principles of Biomedical Ethics*, 6th ed. (Oxford: Oxford University Press).
Châtelet N. (1993) *Trompe l'œil, Voyage au pays de la chirurgie esthétique* (Paris: Belfond).
Clayton L. (1993) *Amphetamines and Other Stimulants* (New York: Rosen Publishing Group).
Coenen C. (ed.) (2009) *Human Enhancement – Study*, Report for the European Parliament (Science Technology Options Assessment).
Daniels N. (2008) *Just Health: Meeting Health Needs Fairly* (New York: Cambridge University Press).
Dean W., Morgenthaler J. and Fowkes S. (1993) *Smart Drugs II* (Petaluma, CA.: Smart Publications).
Elliott C. (2003) *Better than Well: American Medicine Meets the American Dream* (New York: Norton).
Fukuyama F. (2002) *Our Post-Human Future: Consequences of the Biotechnology Revolution* (New York: Picador).

Goffette J. (2006) *Naissance de l'anthropotechnie – De la biomédecine au modelage de l'humain* (Paris: Vrin).
Hagège J.C. (1993) *Séduire! Chimères et réalités de la chirurgie esthétique* (Paris: Albin Michel).
Hottois G. and Missa J.N. (2002) *Species Technica* (Paris: Vrin).
Leriche R. (1966) 'Introduction générale; De la santé à la maladie; La douleur dans les maladies', *Encyclopédie française* (Paris: Comité de l'Encyclopédie Française Éd., vol. 6).
Lidsky T. and Schneider J. (2001) *Brain Candy* (New York: Simon & Schuster).
Missa J.N. and Perbal L. (eds) (2009) *Enhancement: Éthique et philosophie de la médecine d'amélioration* (Paris: Vrin).
Ordre National des Médecins (1988a) 'A propos d'un ouvrage intitulé "300 médicaments pour se surpasser physiquement et intellectuellement"', *Lettre d'information de l'Ordre des Médecins*, n° 10, p. 3.
Ordre National des Médecins (1988b) 'Communiqué à la presse professionnelle et à la presse grand public du 26 août 1988', *Bulletin de l'Ordre des Médecins*, n° 11, p. 273.
Ordre des Médecins (2012) *Code de déontologie médicale* (Paris: Conseil National de l'Ordre). Available from: http://www.conseil-national.medecin.fr/sites/default/files/codedeont.pdf [20 May 2013].
Pelton R. and Clarke-Pelton T. (1989) *Mind Food & Smart Pills* (New York: Doubleday).
President's Council on Bioethics (2003) *Beyond Therapy: Biotechnology and the Pursuit of Happiness. A Report of The President's Council on Bioethics*. Foreword by Leon Kass, M.D., Chairman (New York: Dana Press).
Rothman S. and Rothman D. (2003) *The Pursuit of Perfection: The Promise and Perils of Medical Enhancement* (New York: Pantheon Books).
Sahelian R. (2000) *Mind Boosters: A Guide to Natural Supplements that Enhance your Mind, Memory, and Mood* (New York: St. Martin's Griffin).
Sloterdijk P. (1999) *Regeln für den Menschenpark* (München, Germany: Sührkamp Verlag).
Sloterdijk P. (2000) *La domestication de l'être* (Paris: Mille et Une Nuits).
Souccar T. (1997) *Le guide des nouveaux stimulants* (Paris: Albin Michel).
Zarifian E. (1994) *Des paradis plein la tête* (Paris: Odile Jacob).

3
Why Do We Wish to be Enhanced?
Vincent Menuz

Introduction

Recent progress in the fields of nanotechnology, biotechnology, information technology and cognitive science[1] has led to a growing debate concerning the use of these technologies to increase, among other things, individuals' cognitive and physical capacities, as well as their longevity. In Western countries, new philosophical movements, such as the *transhumanists*,[2] speculate on the possibilities of using these technologies to dramatically 'enhance' human beings (Bostrom, 2005). For such thinkers, humanity is moving towards a *post-human* condition where biotechnologically enhanced humans will bypass the standard Darwinian concept of evolution (Harris, 2007). They argue that ultimately, technological progress will lead these post-humans to reverse the ageing process and to experience extreme longevity, if not immortality (de Grey, 2005; Kurzweil and Grossman, 2009). The idea of such a radical transformation of human beings is not just a curious idea shared by a small group of eccentric thinkers. Nowadays, many widely circulated magazines are voicing transhumanist ideas,[3] while institutes specifically dedicated to the development of technologies to modify human beings are opening their doors.[4] Even if radical technological modifications of human beings are not around the corner (Turner, 2004; Spinney, 2006; Finch, 2010), discussions on the socio-ethical issues related to human enhancement have been going on for some time (Gaylin, 1984, 1990; Kitcher, 1997). However, the broader socio-ethical awakening to this topic occurred with the release, in 2003, of an American report published by the President's Council on Bioethics (President's Council on Bioethics, 2003; Mahowald, 2005). The number of related publications increased dramatically thereafter, with the socio-ethical issues surrounding human enhancement becoming one of the most important topics in bioethics, and, according to some, 'the most significant area of bioethical interest in the last twenty years' (Harris, 2011, p. 102). Interestingly, while discussions have mainly addressed the social and personal consequences of human enhancement on individuals and their offspring (Table 3.1), less attention has been paid to

Table 3.1 Some of the main topics discussed in the literature on ethical issues raised by human enhancement (mainly based on Greely, 2005 and Caplan, 2009)

Topics	Examples of publications
Technological modifications of individuals *would be/would not be* new ways to pursue perfection, which may be considered as vain, selfish and unrewarding	President's Council on Bioethics, (2003), Rothman and Rothman (2004)
The alteration of human biology *would/would not* lead to unfairness, injustice and inequality between enhanced and unenhanced people. Access to the technology *will/will not* be reserved to people who can afford it	Caplan (2009), Elliott (2003), Harris (2007), Mauron (2005) Schermer (2008b), Van Hilvoorde and Landeweerd (2010)
Increased performance achieved by enhancement technologies *would/would not* be 'authentic', because it *will/will not* separate individuals from who they really are and from how the world really is, as well as undermine the notion of effort as being required to achieve something	Bolt (2007), DeGrazia (2005), Safdar et al. (2011), Sandel (2004), Schermer (2008a)
Improvement of human beings by artificial means *would/would not* represent a threat to human nature	Bostrom (2005), Buchanan (2009), Daniels (2009), Elliott (2003), Sandel (2004)
The outcomes of technological interventions on human beings *are/are not* uncertain and the safety of their use *can/cannot* be taken for granted	Drabiak-Syed (2011), Greely (2005), Hotze et al. (2011)
Human enhancement *would/would not* lead to active or passive forms of coercion	Appel (2008), Sandel (2004), Warren et al. (2009)
Enhancing individuals *would/would not* undermine solidarity between individuals and increase personal responsibility	Lev (2011), Sandel (2004)
Designing children *would/would not* (i) undermine and distort the role of parents and (ii) project the shadow of eugenics	(i) Kamm (2009), Sandel (2004). (ii) Kitcher (1997), Koch (2010), Sandel (2004), Savulescu and Kahane (2009), Sparrow (2011)
Selecting the best possible child *should/should not* be mandatory, in order to increase their chances of well-being in life	Savulescu (2009), Sparrow (2011)

the nature of the personal motivations that may drive individuals to undergo enhancing technological interventions with a view to reaching a given objective. This is intriguing because there is a necessary link between the use of general or specific technological interventions and the desires – or motivations – that may drive individuals to use them. In addition, we recently proposed a definition of human enhancement that is built around individual perceptions (Menuz et al., 2013). In such a perspective, exploring what may motivate individuals to undergo technological interventions is an important issue to consider. As recently put forward by Carl Elliott, such motivations are influenced by social forces (Elliott, 2011). Therefore, the contexts in which such motivations arise are also crucial elements to explore.

Based on an analysis of the scientific literature addressing socio-ethical issues related to human enhancement, the first part of this chapter establishes an inventory of objectives which individuals seek to reach through technological interventions. The second part explores how socio-cultural frameworks might influence individuals to consider certain objectives as something essential towards which one should reach. While none of the authors surveyed identified motivations other than those defined by the expected outcomes of technological interventions, those very motivations to undergo a technological intervention could be strongly influenced by socio-cultural forces that shape the values and beliefs that materialise as individual objectives to personal enhancement.

Identification of personal motivations to be enhanced

In order to better understand how individuals' motivations to be technologically enhanced are portrayed in scientific publications, we systematically analysed 109 journal articles on socio-ethical issues related to human enhancement published between 2006 and 2011 (see 'Methods' section for details). This screening empirically confirmed that issues related to individuals' personal motivations to undergo technological modification have been largely overlooked. Indeed, only three articles out of 109 addressed issues related to individuals' motivations (Buyx, 2008; Hotze et al., 2011; Lamkin, 2011). While Buyx and Hotze et al. tackled questions related to physicians' attitudes towards patient requests for biotechnological modifications (for example, increasing working skills or athletic performances), none of them explored the situations that had led their patients to express such requests. Only one article focused, to a certain extent, on the personal motivations that could drive individuals to undergo technological interventions as well as some of the factors that could have influenced – or given rise to – such motivations. In that article, Matt Lamkin developed striking arguments to shed light on the social pressures impelling individuals to use different kinds of technologies for lightening their skin. According to Lamkin, a 'darker skin correlates with reduced opportunities and negative outcomes' (Lamkin, 2011, p. 185) and skin lightening may be considered a human enhancement for dark-skinned individuals. Importantly, the paper highlighted two key points. First, the result of a technological intervention may be considered a human enhancement by some, but not by others (Menuz et al., 2013). Indeed, skin lightening may not be perceived as a human enhancement for white people (who may even prefer skin-darkening for aesthetic reasons). Second, it illustrates some of the social pressures at play behind the wish for human enhancement, a point that will be discussed further on.

Our analysis shows that, in a vast majority of publications, authors address the features individuals wished to modify (for example, strength, cognition, rate of ageing) as well as the objectives (for example, winning a race, increasing productivity at work, living longer) they pursued. They do not focus on the potential factors influencing individual motivations to be enhanced. In other words, authors do not attempt to identify other motivations or reasons than those defined by the expected outcomes of technological interventions on body or mind and their attendant results (for example, individuals wanting to increase strength to win a race, or heighten alertness to boost productivity at work or decelerating their aging rate in order to extend their lifespan). However, the expected outcomes of technological interventions are likely (and logically) part of the motivations that lead individuals to desire them. Therefore, the objectives of specific technological interventions such as those provided by authors may constitute a first and necessary step in an attempt to address more about the potential personal motivations of individuals to be enhanced. We therefore used a general inductive approach (Blais and Martineau, 2006) to identify and organise the different kinds of objectives provided by authors around four main categories: individuals' wishes for (1) interventions aimed at adapting to the environment, (2) interventions aimed at fighting disease, ageing and death, (3) interventions on existing or future children and (4) interventions aimed at increasing happiness/well-being (Table 3.2).

At this stage, a few remarks are in order. First, interventions aimed at optimising interactions amongst individuals or with their social environment (that is increasing social acceptability, improving social,

Table 3.2 Technological interventions and their objectives as described by authors. The interventions and their objectives are classified into four main categories, which in turn are organised into more specific sub-categories (first columns)

Sub-categories	Examples of technological interventions and their objectives
Category 1. Interventions aimed at adapting to the environment	
Increasing social skills through psychological or cognitive modifications	– Reducing social anxiety, hatefulness, aggressiveness, jealousy, shyness – Improving sociability, agreeableness, conscientiousness, assertiveness, self-confidence, honesty, fairness – Improving morality (for example, ability to make a difference between right and wrong, avoiding inflicting pain or suffering on others, altruism, sensitivity, consideration and so on) – Improving control of emotions, moods, impulsive behaviours – Improving social intelligence
Increasing social acceptability through physical modifications	– Lightening skin for social acceptability – Modifying morphologies (for example, augmentation of breast size, choosing height), cosmetic surgery (for example, face lifting, breast implants, wrinkle smoothing)
Increasing working skills through psychological, cognitive or physical modifications	– Reducing fatigue, stress – Improving calmness – Improving strength, speed, stamina, reaction speed, morphologies, productivity, sensory and motor abilities (for example, steady surgeon's hands) – Reducing fear in people with dangerous jobs – Increasing productivity – Improving concentration, memory storage and working memories, attention, focus, performance on various complex motor learning tasks, proofreading, numerical ability, learning skills (for example, language learning), verbal fluency, abstract reasoning, spatial cognition, resistance to sleep deprivation, alertness and wakefulness
Increasing learning skills through cognitive modifications	– Improving learning speed, concentration, memory storage and working memories, attention, focus, proofreading, numerical abilities, learning skills, abstract reasoning, spatial cognition, language learning, alertness and wakefulness

Increasing athletic skills through psychological, cognitive or physical modifications	– Reducing fatigue, stress – Improving calmness – Improving strength, speed, stamina, reaction speed, morphologies, sensory and motor abilities – Improving concentration, attention, focus, performance on various complex motor learning tasks, abstract reasoning, spatial cognition, alertness and wakefulness
Increasing sexual skills through psychological or physical modifications	– Improving pleasure and arousal – Correcting erectile dysfunctions – Increasing/reducing breast size – Using hormonal contraception
Increasing seduction skills/appearance through physical modifications	– Improving physical appearance with cosmetic surgery – Modifying morphologies (for example, augmentation of breast size, choosing height) – Maintaining/recovering a youthful appearance – Reducing the need for exercise to remain physically fit – Choosing colouring, lightening/darkening skin
Increasing fighting skills (for soldiers) through psychological, cognitive or physical modifications	– Reducing fear in combatants – Increasing aggressiveness – Erasing memories of traumatic events – Improving strength, speed, stamina, reaction speed, morphologies (for example, height, muscle hypertrophy), talents, sensory and motor abilities – Identifying gene markers to distinguish friend from foe between combatants – Improving combatants' cognitive capacities (for example, alertness), resistance to sleep deprivation, concentration, attention, focus, ease of skill acquisition (for example, language learning), abstract reasoning, spatial cognition, resistance to sleep deprivation, alertness and wakefulness
Fitting into an artificial, fast-changing environment	– Taking charge and modifying human biological development and evolution in order to better fit into the artificial environment human beings are continuously creating

Continued

Table 3.2 Continued

Sub-categories	Examples of technological interventions and their objectives
Category 2. Interventions aimed at fighting disease, ageing and death	
Health optimisation	– Improving resistance to disease, manipulating the immune system (for example, vaccination), treatment of genetic risk factors – Improving lifestyle, health-wise – Using diagnostic tests – Determining genetic transmission of disease – Optimising individual lifestyle – Controlling impulses to limit unhealthy food intake
Fighting ageing	– Retarding ageing and age-related medical conditions, increasing health-span – Reversing ageing – Eliminating ageing (for example, unlimited youth and vigour, youthful body functions)
Extending lifespan	– Helping people living longer (for example, increasing lifespan) – Seeking indefinite health span, quest for immortality
Category 3. Interventions aimed at modifying existing or future children	
Altering existing or future children	– Choosing for certain features/properties of the child – Selecting embryos with pre-implantation genetic diagnosis (PGD) – Selecting genetic traits (disease and non-disease variations) – Selecting embryos for sex (for example, for family balancing) – Selecting a child with 'superior' genes that would allow it to have a life expectancy of 100 instead of 75 years.
Category 4. Interventions aimed at increasing happiness/well-being	
Increasing happiness/well-being through psychological modifications	– Erasing memories of shameful conduct, of traumatic events – Reducing guilt, depression, anxiety, grief, fear, stress – Improving calmness, capacities for pleasure, serenity, love, artistic appreciation – Experiencing novel states of consciousness

working, learning, athletic, fighting or sexual skills, and fitting into an artificial, rapidly changing environment) have been put together in the category 'interventions aimed at adapting to the environment' (Table 3.2, Category 1). The category 'interventions aimed at fighting disease, ageing and death' (Table 3.2, Category 2) comprises interventions that are related to the biological nature of human beings, such as health, ageing and length of life (that is health optimisation, fighting ageing and extending lifespan).

Second, individuals opting for 'interventions aimed at modifying existing or future children' (Table 3.2, Category 3) may actually be pursuing any of the objectives that were identified and classified elsewhere in Table 3.2. For instance, parents may want to conceive a child with strong muscles in order to increase their athletic skills. Similarly, it is difficult to deal with an objective generally described as the increase of happiness (Table 3.2, Category 4), as such an objective may actually be reached with the achievement of other objectives presented in Table 3.2. For instance, an individual may wish to increase their happiness by increasing working or athletic skills as well as by extending their lifespan. Ultimately, we may even wonder whether increasing happiness does not constitute the general objective of any technological interventions on human beings (Elliott, 2011). An attempt to synthesise or fully grasp these issues would go beyond the scope of this chapter and we will thus mainly focus on categories 1 and 2 as described above.

Third, interestingly, the four kinds of interventions that have been identified through our analysis of literature (that is interventions on existing or future children, those aimed at increasing happiness/wellbeing, those aimed at better adapting to the environment and interventions aimed at fighting disease, ageing and death) recall the classification used in the American President's Council on Bioethics report, *Beyond Therapy*, published in 2003 (that is 'Better children', 'Happy souls', 'Superior performance' and 'Ageless bodies'). One wonders whether this reflects the influence this report may have had on socio-ethical discussions related to human enhancement.

Fourth, importantly, the list of objectives reported in Table 3.2 only reflects the literature addressing socio-ethical issues related to human enhancement and it is not possible to determine whether these issues correspond to the objectives that individuals are pursuing in their quest for enhancement. In other words, this table does not provide an exhaustive list of possible individual motivations to be enhanced. However, Table 3.2 may constitute a useful baseline to initiate a reflection on individuals' motivations to be technologically modified. As a second step, and in order to grasp the complexity and diversity of individual motivations, it is then necessary to ask why individuals would actually want to reach the objectives described in this table. This question cannot be answered without considering the influence of the socio-cultural frameworks in which individuals exist.

Influences of socio-cultural frameworks on individuals' motivations to be enhanced

Prior to any biotechnological intervention, there is some appeal for it. Recently, Carl Elliott proposed that such an appeal is rooted in social forces related to a need for social recognition, a quest for authenticity and a willingness to consider human beings as 'objects of potential control' (Elliott, 2011, p. 372). Put another way, while the very concept of human enhancement is subject to socio-cultural influences, such as political and social norms, environmental factors, different forms of passive coercion and statistically defined attributes, as well as personal considerations (Menuz et al., 2013), it would stand to reason that similar influences could also shape an individual's motivations to seek technological intervention in order to be enhanced (Elliott, 2011). As a consequence, it does not suffice to identify individuals' stated objectives to be enhanced; one should also identify the socio-cultural forces (the exact nature of the appeal) that underlie such objectives.

Socio-cultural influences on motivations for interventions aimed at adapting to the environment

The quest for performance

In many societies, individuals – regardless of sex, age or social status – are continuously driven to performance which is rooted, at least in part, in an individual's wish, or the necessity, for social recognition or social survival (Laure, 2002). Such a need for performance acts as a social force that may lead individuals to consider technological intervention as a response to such pressure. For instance, in order to build an academic career, junior scholars have to rather quickly accumulate a series of excellent, peer-reviewed publications in high impact factor journals, while senior academics have to continually write impressive grant proposals and articles in order to fund their research as well as show their peers how well they can still perform. In institutional culture, this phenomenon is captured by the maxim 'publish or perish', which highlights that 'professional success depends almost exclusively on our publication record' (Ladle et al., 2007, p. 25). The risk of 'perishing' acts as a strong driving force, leading some academics to take nootropics – such as the molecule *Modafinil*, for instance – that may increase their alertness and productivity, as well as help them to sustain hard thinking (Sahakian and Morein-Zamir, 2007). Students can be confronted with pressure similar to that facing academics (for example, increasing alertness) and may likewise respond by relying on so-called 'cognitive enhancers' (Babcock and Byrne, 2000). Even if the use of such 'cognitive enhancers' among students has been reported as being minor (Outram, 2010), it illustrates a tendency to use technological modification as a response to social pressure.

On-the-job performance is central to all professions and any worker may wish to use technological possibilities in order to increase – or maintain – their performance level. However, some professions – such as surgeon, truck driver or airline pilot – involve a significantly higher degree of responsibility and, accordingly, a higher degree of performance is expected from those who practice it (Warren et al., 2009). For instance, it seems reasonable for a patient to expect from a surgeon that the latter have, say, perfect vision for precise tasks, as well as acute cognitive capacities. Such a patient's expectations, related to a surgeon's performance, might passively coerce a certain number of surgeons to undergo technological modification (Babcock and Byrne, 2000). Moreover, if the use of nootropics, for instance, becomes common among professionals, a passive coercion to use them may be exacerbated by peer pressure to adopt them. Ultimately, regulations could be enacted in order to mandate some forms of technological modification (Appel, 2008). Indeed, 'once the use of an enhancement technology becomes widely accepted, it paves the way for changing social institutions in a way that drives the demand for the technology even further' (Elliott, 2011, p. 367). In other words, over time, some technological modifications may become either expected or mandatory, becoming part of the very culture of a given institution. Such a point is well illustrated by professional cycling, in which just such a culture of technological modification already exists. According to some authors, professional cyclists have developed a culture of doping that is specific to their sport (Lentillon-Kaestner and Brissonneau, 2009; Schneider, 2006). This culture is embedded in a complex context involving both a willingness to perform (Lentillon-Kaestner, 2008) and peer pressure (Schneider, 2006). Professional cyclists consider doping as a 'normal' part of their athletic activity and, as a consequence, their wish to take an ergogenic substance is a wish to maintain this 'normality'.

The quest for performance is also an undeniable hallmark of the army, whose interest is certainly to enhance soldiers' physical, psychological and cognitive capacities. For instance, in the United States of America, the Pentagon's Defense Advanced Research Project (DARPA) aims 'to exploit the life sciences to make the individual warfighter stronger, more alert, more endurant, and better able to heal' (from Moreno, 2004, quoting a US Defense Department paper headed 'Enhanced Human Performance, Strategic Plan', DARPA, February 2003). The will to increase soldiers' capacities through technological modification is rooted, at least in part, in national security objectives as well as in the efforts to minimise casualties among fighters. However, soldiers have to obey orders and they may be compelled to accept some technological modifications. As a consequence, their motivations are irrelevant, and the army may represent one of the most striking examples of active coercion forcing individuals to undergo technological modifications. In addition, it shows that, in certain circumstances, the objectives such as those described in Table 3.2 could be objectives that are pursued by an agent (in this case the military) other than those who undergo the interventions *per se* (that is the soldiers). This is certainly a crucial element that should be taken into account when addressing the potential motivations of an individual to be enhanced.

The quest for performance also intrudes into individuals' intimate lives. Indeed, sexuality is also subject to performance standards and associated pressures. The search for a potion aiming at increasing sexual performance has a long tradition and unscrupulous profiteers have tried to take advantage of this human desire, as well as of the anxiety resulting from any – real or perceived – deficiency in this regard (McMahon, 2010). The exploitation of such an anxiety has even spilled over into the public health agenda, as illustrated by the recent use of the fear of impotency as a way of persuading men to quit smoking. It is believed that 'if widespread publicity came to forge a natural apposition between the cigarette and a vision of a flaccid, dejected penis, the semiotics of smoking may change radically' (Chapman, 2006, p. 74). To put it another way, such a campaign is based on the idea that the quest for sexual performance can exert a powerful influence on individual behaviour. More recently, sexual activity has been directly related to health with the publication of scientific studies demonstrating a possible link between sexual activity and health benefits, such as, for instance, decreasing the risk of cardiovascular disease (Ebrahim et al., 2002) or the decrease of depressive symptoms (Gallup et al., 2002). The mass media have widely reported these discoveries (Gupta, 2011), thereby establishing a science-based justification for a 'sex-positive attitude', implying that individuals not having frequent or 'effective' sexual activity suffer from a pathologic condition (Gupta, 2011). The media have indeed played a key role in normalising current sexual behaviours (Pinkleton et al., 2012). They frame a kind of normative sexual activity (Gupta, 2011) that may increase, if not create, anxiety regarding sexual performance in an individual. As a consequence, in many societies, significant social pressure linked to sexuality is at work and may constitute a form of passive coercion that can undoubtedly influence individuals' motivations to turn to transient – or permanent – biotechnological modifications.

The need for social adaptation

As shown above, socio-cultural frameworks may influence, and even force, individuals to consider technological modification as a solution to meet some of the challenges raised by a growing and pressing demand for performance. It may be important to remember that individuals bear a responsibility in creating such socio-cultural frameworks. By conveying normative views, the public recasts socio-cultural frameworks which determine social norms – or social standards – that individuals would do well to follow in order to escape the risks of stigmatisation and/or discrimination (Raphael, 2010). For instance, aesthetic standards, mainly conveyed by the media (Haas et al., 2008), have taken on an imperative urgency that pressures people to do their best to look more attractive, thinner or younger. One consequence is that the recourse to cosmetic plastic surgery has surged, one of a number of solutions adopted in response to these normative standards (Haas et al., 2008; Raphael, 2010; Wijsbek, 2000). Similarly, social norms may render some behavioural traits such as shyness, social anxiety, anger, jealousy, selfishness or, even, 'lack of morality', to be highly undesirable. Such norms may also exert pressure on individuals to use technological interventions the better to comply with the said norm (Greely, 2012). Many forms of discrimination and stigmatisation are rooted in social norms. Indeed, racism, sexism and ageism are all examples where stigmatised individuals are unable to fit socio-cultural standards set by a particular society. The pressure that such groups face may lead some individuals to consider technological modifications as a solution to fix, or at least diminish, stigmatisation and its consequences (Lamkin, 2011).

Importantly, while media and the public may, in part, bear a responsibility for conveying socio-cultural norms, some academics have recently started to give a normative dimension to some objectives people should follow, given that technological interventions could become available.

For instance, according to Julian Savulescu, the freedom of human beings is constrained by biological and psychological barriers, and removing such constraints through technological means should be 'a moral obligation', as it might help us to act 'morally' (Savulescu, 2010, p. 04.2). It has also been proposed that technological interventions that 'can prompt solidarity should be candidates for state support' (Lev, 2011, p. 181). More radically, according to some, technological modifications that may 'reduce harm to others should be obligatory. The State ought to consider enforcing such obligations' (Lev, 2011, p. 181). Finally, while discussions on the moral obligation to bring into this world only those children with the best possible chances in life have been going on for some time (Harris, 2007; Savulescu and Kahane, 2009; Sparrow, 2011), S. Matthew Liao has recently explored the consequences related to the duty for parents to love their children, given that a 'love pill' could become available (Liao, 2011). Such perceived or proposed obligations and duties would seem to legitimise a form of technological interventionism with regard to individuals. Some technological modifications are envisioned as eventually becoming mandatory, with individuals left no choice but to submit to them.

The pressure to adapt to an artificial fast changing environment
Modifying the environment to fulfil certain needs is not specifically or exclusively human, but a typical trait of animal functioning. Birds build nests, beavers build dams, and ants and bees build complicated structures which accommodate their complex societies. However, human beings are the only species to have created a totally artificial environment in order to serve its needs (Canto-Sperber, 2008). Nowadays, many people are clearly living in an artificial, quickly changing environment in which it has become difficult to make the most of every opportunity on offer. For instance, it is becoming difficult, if not impossible, to follow the perpetual flow of daily information as well as the continuous evolution of new possibilities offered by computers. Just think about continuous software updating, virtual social networks as well as new devices such as smartphones or tablet computers. Social pressure – from relatives, friends, employers or colleagues – may compel individuals to attempt to continuously exploit all these possibilities and seek to adapt to this constant evolution. Not doing so may lead to social isolation and ultimately to stigmatisation and discrimination. For some authors, the human species is no longer adapted to modern conditions and technological modifications should be seen as a necessary means to 'enhance our evolution' and better fit into such an artificial environment (Harris,

2007; Powell and Buchanan, 2011). Therefore, the artificial environment itself may have a direct impact on individuals' motivations to be technologically modified.

To summarise this section, it appears that individual motivations to undergo technological interventions aimed at adapting to the environment cannot be addressed properly in terms of objectives only, without considering the socio-cultural factors that may influence the individuals. As developed below, the same is certainly true for the motivations of individuals who are turning to technological modifications aimed at fighting disease, ageing and death.

Socio-cultural influences on motivations for interventions aimed at fighting disease, ageing and death

An imperative to fight ageing and death?

Initially, defeating disease, ageing and death was central to the aspirations of the founders of modern science (Amundsen, 1978; Callahan, 2000; President's Council on Bioethics, 2003). During the last century, progress in medical practice partially met these expectations. New ways to fight infectious diseases were developed (mainly thanks to hygiene, vaccines and antibiotics) allowing more people to live longer (Fries, 1983; Vijg and Campisi, 2008). As a consequence, in Western countries for example, the life expectancy of people increased by about 50 per cent during the last century (Wilmoth, 2000). However, as Aubrey de Grey claims, 'ageing is bad for you'.[5] Indeed, ageing is accompanied by a progressive degeneration of the body as well as a greater likelihood of physical and/or mental illness (Blumenthal, 2003; de Grey, 2007; Gems, 2011; Juengst et al., 2003; President's Council on Bioethics, 2003; Wimo and Prince, 2010). Therefore, the dramatic increase in life expectancy in the 20th century marked an epidemiological transition: 'increasingly, acute infectious diseases declined and mortality risks shifted to older ages, in conjunction with the rise of chronic degenerative diseases' (Barazzetti and Reichlin, 2011, p.1 p. E1). To put it in another way, a significant increase in life expectancy is coupled with progressive, age-related, debilitating senescence, consequences that may imperil the achievements of modern medical science. Therefore, according to some commentators, it is now imperative to launch a 'mission' against age-related diseases and, by the same token, against ageing itself (de Grey, 2007; Farrelly, 2010; Gems, 2011; Mallia, 2010; Rae et al., 2010).

According to Aubrey de Grey and Collin Farrelly, age-related diseases are a leading cause of death, killing 'roughly 100,000 people every day, worldwide' (de Grey, 2007, p. 2; Farrelly, 2010). This leads de Grey to argue that most industrialised societies are victims of a 'pro-ageing trance', which prevents them from understanding the imperative of concretely starting a fight against ageing (de Grey, 2007). Until recently, ageing was considered a 'normal' part of the life cycle and fighting ageing, just as extending lifespan, were both considered as neither possible nor desirable (an attitude called 'apologism', Gruman, 1966). However, such positions came to be increasingly challenged with the discovery that ageing and longevity both exhibited a certain plasticity. For instance, in model organisms, mutations in some specific metabolic pathways influence both the rate of ageing and the length of lifespan (Kenyon, 2010). Similarly, non-genetic interventions such as caloric restriction – that is the reduction of daily caloric intake – have been shown to be an effective way both to postpone age-related symptoms and to extend lifespan in model organisms (Fontana et al., 2010). According to Farrelly, 'of the various goals a rational society seeks to realise, that of preventing harm by reducing a population's risk of morbidity and mortality is amongst the most paramount goals' (Farrelly, 2010, p. 3). As a consequence, given that ageing is a leading cause of disease and death (de Grey, 2007; Farrelly, 2010; President's Council on Bioethics, 2003), and that 'scientific research has demonstrated that ageing is not immutable' (Farrelly, 2010, p. 4), Farrelly has recently proposed that research aimed at retarding human ageing should become an imperative (Farrelly, 2010).

As to the President's Council on Bioethics (President's Council on Bioethics, 2003), it suggested that the quests to decelerate ageing and extend lifespan both shared the goal of conquering death itself, at least implicitly, and that there may indeed be a link between 'the open pursuit of ageless bodies and the secret longing to overcome death' (President's Council on Bioethics, 2003, p. 183). Several authors have more broadly pointed out that fighting death may have implicitly become one of modern science's goals (Amundsen, 1978; Callahan, 2000, 2002; Kiefer, 2010). In this regard, it is important to distinguish between fighting '*premature* death' and fighting 'death *per se*'. On the one hand, fighting *premature* death refers to an ideal of longevity[6] – or the 'ideal of a finite vision' (Callahan, 2002, p. 889) – which is a normative view of the human lifespan. For instance, Callahan defines such an ideal of longevity as a lifespan that would last 'not much longer than that now enjoyed by the healthiest countries in the world, around 80 years of age' (Callahan, 2002, p. 889). In other words, fighting *premature* death would consist of seeking solutions to conditions that may prevent individuals from reaching such an ideal age and where 'extra years [that is above the ideal of longevity] would not have been deliberately aimed at' (Callahan, 2002, p. 889). On the other hand, fighting death *per se* refers to a quest for immortality (President's Council on Bioethics, 2003, p. 183, in a footnote citing William Hazeltine), the 'Holy Grail of human enhancement' (Harris, 2007, p. 59). In this sense, the President's Council on Bioethics (2003) suggested that 'the quest for any age-retardation suggests no inherent stopping point, and therefore, in the extreme case, it is difficult to distinguish it from a quest for endless life' (President's Council on Bioethics, 2003, p. 182–3). Postponing death may therefore still constitute one of the major goals of contemporary science (Callahan, 2000; Harris, 2007; Hayry, 2011; Hazeltine, 2000; Kiefer, 2010), which reinforces the idea of a medical imperative to avoid death above all (Amundsen, 1978; Scheunemann and White, 2011) and the societal assumption that 'sudden death is *premature* death, regardless of age (Amundsen, 1978, p. 11; Kaufman et al., 2011).

A context that promotes both the medical imperative and the societal assumption certainly contributes to legitimate technological interventions that aim to decelerate ageing or fight death. Yet, it is also crucial to consider the factors that have led to the development of such a context in order to address individuals' motivations to fight ageing and death. Below, we focus on the following factors: the fear of disease and death, ageism and the impact of mass media.

The fear of disease, ageing and death as factors in the motivations to fight ageing and death

Human beings have evolved into fearful creatures (Boyer and Bergstrom, 2011). Indeed, according to Ernest Becker in *The Denial of Death*, 'early men who were most afraid were those who were most realistic about their situation in nature, and they passed on to their offspring a realism that had a high survival value' (Becker, 1973, p. 17). While this survival value has played an essential part in the Darwinian evolution of humans, it has also led us to become 'hyperanxious animals' (Becker, 1973, p. 17). He states that, among other anxieties, death anxiety, 'the idea of death, the fear of it, haunts the human animal like nothing else' (Becker, 1973, p. ix). Basically, death anxiety is a 'state in which an individual experiences apprehension, worry, or fear related to death and dying' (Lehto and Stein, 2009, p. 31 citing Carpenito-Moyet, 2008). Briefly, death anxiety comes from awareness of death, making humans the only living beings to live with a consciousness of their own mortality, and, when death awareness increases, death anxiety increases as well (Lehto and Stein, 2009). Importantly, while considered a universal phenomenon, fear of death is strongly repressed by individuals in order to let them function normally (Becker, 1973). Therefore, death anxiety is typically outside conscious experience (Lehto and Stein, 2009) and influences many human behaviours from outside their field of awareness (Becker, 1973). Mainly based on Ernest Becker's work on the denial of death (1973), some contemporary psychologists have developed a Terror Management Theory (TMT), which proposes that individuals seek to reduce their death anxiety by endorsing cultural beliefs that help them to cope with the knowledge of their own mortality (Pyszczynski et al., 1999). According to this view, many cultural beliefs, such as the association of youthful appearance with positive characteristics (Benton et al., 2007) for instance, could be rooted in death anxiety. Such beliefs would influence individuals' motivations to undergo technological modifications with the unconscious objective of keeping the terror of death at bay.

Death anxiety has also been associated with 'ageing anxiety', which can arise when individuals experience a combined concern for, and anticipation of, personal losses related to the ageing process itself (Lasher and Faulkender, 1993). Lasher and Faulkender have developed an Ageing Anxiety Scale (AAS), which distinguishes between different components of ageing anxiety such as, among others, the fear of the ageing process itself. It turns out that death awareness increases death anxiety (Lehto and Stein, 2009). Therefore, the ageing process is accompanied by death's ever greater proximity in time and an increasing awareness of the certainty of one's death, with the likelihood of death increasing with each passing year, until it approaches near-absolute certainty in its clear imminence. Thus ageing itself most likely leads to increased death anxiety (Benton et al., 2007), which would quite reasonably, in turn, affect individuals' motivations to use technological interventions aiming at decelerating their ageing or extending their lifespan.

In turn, ageing anxiety is certainly fed by the fear of losing one's health. Health has become a widespread preoccupation in contemporary societies. In a striking article, Robert Crawford traces this contemporary health obsession to Enlightenment ideals of the 1700s (Crawford, 2006). According to Crawford, since the 1980s, two trends have appeared in contemporary Western societies. Governmental policies, medical discourse and the mass media created, on the one hand, a strong incentive for citizens to embark on a quest for better health, with the risk of stigmatising those who did not adapt their behaviour to this quest. On the other hand, an emphasis has been put on the constant danger of losing one's health. This last trend greatly contributed to making individuals increasingly anxious about health issues (Crawford, 2006). A famous line from a Woody Allen film – 'the most beautiful words in the English language are not *I love you*, but *it's benign*'[7] – illustrates this anxiety well (Crawford, 2006), an anxiety which is leading contemporary health practice to 'a logic of survival [where] individuals must do what they can to protect themselves from harm' (Crawford, 2006, p. 419). Just such a survival logic is likely at work when the harm comes from ageing and death.

Altogether, the fear of disease, ageing and death are crucial factors when it comes to addressing individual motivations to use technological interventions aimed at decelerating ageing or extending lifespan.

Ageism as a factor in motivations to fight ageing and death

Ageism is a form of discrimination based on age (Bodner et al., 2012). It is possible to distinguish between individual and institutional forms of ageism. While the former refers, for instance, to the 'avoidance of contact with older people, age denial, ageist humour, patronising and negative attitudes and stereotypes about older adults', the latter refers, among others, to 'discrimination in housing employment, mandatory retirement, public policy, and inappropriate care in institutional settings' (Bodner et al., 2012, p. 1). Importantly, even if there are cultural differences between societies, ageism has been described as a worldwide, unavoidable form of discrimination (Martens et al., 2005). It is suggested that ageist attitudes among young people could be explained, at least in part, by TMT (Bodner, 2009). As briefly introduced above, older persons tend to be associated with death. Therefore, according to TMT, 'younger persons may adopt ageist attitudes and behaviours to distance themselves from older people in order to deny the reality that they, too, will eventually become part of that [group] and die' (Bodner, 2009, p. 1007). In other words, in young people, ageism may be seen as a way of escaping the death anxiety that older people may induce or intensify in them. And yet, it is common for older people to develop ageism against people of their own age (Bodner, 2009). However, unlike in young people, fear of death does not seem to play a role in the ageism that the elderly may develop between them. According to Ehud Bodner, 'when an individual is a member of a group that is often perceived as having a lower status (i.e., being old), this individual will adopt one or more strategies to maintain a positive self-identity' (Bodner, 2009, p. 1005). Accordingly, some elderly will tend to distance themselves and display ageist attitudes towards other members of their own age group. Ageism may therefore represent a strong but also complex factor that influences individuals' motivations to undergo technological modifications aimed at decelerating ageing, or at least diminishing the scars of time.

Mass media as a factor in the motivations to fight ageing and death

As the will to fight ageing and to prolong life is a long-held human dream (Gruman, 1966), it is not surprising that the mass media have started communicating the information that science is on the verge of proposing solutions to counter ageing (Helmuth, 2005) as well as ways of extending the human lifespan. However, they often do so with much hype (de Grey, 2010). In France, for instance, the cover of *Le Figaro Magazine* (January 8th, 2011, n°1576) showed the close-up of a cute baby with the caption 'Will he live 130 years? How science has revolutionised our life expectancy' (my translation). Similarly, the front cover of the well-known American news magazine *Time* dated February 21st, 2011, showed the back of a bald-head connected to a cable with the words '2045, the year man becomes immortal'. While such claims are clearly premature (Finch, 2010), such hype shapes a mass culture in which individuals are convinced that fighting ageing and extending lifespan is not only feasible, but also desirable (Mykytyn, 2006). Recently, this phenomenon has been reinforced by the publication of popular science books, written by physicians, stressing the inevitable success of attempts to delay ageing and to extend lifespan in the near future (for example, Alexandre, 2011). Scientific articles, written by gerontologists, have urged for the translation of some lab discoveries into interventions against ageing in humans (for example, Rae et al., 2010). The scientific authority of gerontologists and physicians may contribute to unrealistic or exaggerated expectations that reinforce the idea that it is legitimate to fight ageing and then to turn to technological interventions that may contribute to reach such a goal.

In summary, similarly to technological interventions aimed at better adapting individuals to the environment, individual motivations to undergo technological interventions aimed at fighting disease, ageing and death cannot be addressed solely or strictly in terms of the objectives individuals set for themselves. The wider and deeper socio-cultural forces that influence them should be taken much more fully into consideration than they have been up to now.

Conclusion

This analysis of the literature shows that authors describe the objectives of technological interventions aimed at enhancing human beings without focusing and most often even without addressing the many factors that may influence individuals' motivations to undergo such interventions. Yet, as shown in this chapter, these motivations are rooted in various, and also complex, socio-cultural frameworks. In order to address the ethical issues at stake in human enhancement, it is necessary to understand the complexity of the factors that may lead individuals to desire such an enhancement. The pressures, incentives and obligations to increase individual performance or social acceptability as well as to adapt to a rapidly changing artificial environment must be taken into consideration. Similarly, individuals are under strong socio-cultural pressures that influence their perspectives on health, age, lifespan and death, and that may shape their desire to turn to technological interventions aimed at improving their health and extending their lifespan. Any person seeking to understand the complexity and the controversies surrounding human enhancement must be aware of the driving socio-cultural factors behind individual motivations to seek such technological interventions. Ultimately, we may come to wonder whether individuals who want to undergo these interventions actually want to reach the same objectives as the ones described by commentators or whether they are mainly considering such interventions in order to comply with the many pressures and incentives present in our societies. Providing an answer to this question is of the utmost importance in a debate about the ethical conundrums raised by human enhancement.

Methods

In order to understand what might lead individuals to undergo biotechnological modification, we compiled a corpus of publications addressing ethical issues related to 'human enhancement' using the interface OvidMEDLINE,[8] which provides powerful tools for detailed research of publications on the MEDLINE[9] database. *OvidMEDLINE(R)-In-Process & Other Non-Indexed Citations* (that is, segment PREM) and *OvidMEDLINE(R) 1948 to Present with Daily Update* (that is, segment MESD) were used. The research strategy can be summarised as such: {*Advanced search*: [Main heading: Biotechnological enhancement/sub-heading: Ethics] OR *Advanced search:* [Main heading: Genetic enhancement/sub-heading: Ethics]} AND {*Search field:* [Journal article]}. Publications were selected on the basis of their language (English) and their date of publication (2006–2011, last search December 15th, 2011). We decided to concentrate on these specific years because the number of publications addressing ethical issues linked to human enhancement has been increasing since 2006. This may be explained by the release of the President's Council on Bioethics report in 2003, which specifically addressed socio-ethical issues linked to human enhancement (President's Council on Bioethics, 2003). We found 181 papers indexed as 'journal article' by MEDLINE. After analysis, 109 Journal articles were found (59.9%), as well as 48 comments (26.4%, that is 3 *Commentaries*, 39 *Responses to Target Article*, 4 *Responses to Open Peer Commentaries*, 1 *Introduction to Special Issue* and 2 *Book reviews*), 4 editorials (2.2%), 11 were not pertinent (6%), 9 were not accessible from the library of the University of Montreal (5%) and 1 article was not found (0.5%).

Table 3.2 was generated using an inductive approach (Blais and Martineau, 2006). The 109 journal articles were analysed in order to determine categories based on (i) the themes addressed in the publications (for example, sports enhancement, student neuroenhancement) and (ii) the examples used by authors to illustrate the objectives of interventions in terms of technological modification on individuals (for example, making people stronger, smarter, happier). Finally, these categories were compared, organised and classified according to their similarities or differences.

Acknowledgments

I am greatly indebted to Professor Béatrice Godard and Thierry Hurlimann for their helpful and constructive comments on earlier versions of this manuscript. The project underlying this chapter was supported by the Swiss National Science Foundation as well as by the Apogée-Net/CanGèneTest network.

Notes

1. These technologies are usually referred to as 'NBICs' and are often described as 'emerging and converging technologies' (Roco and Bainbridge, 2003).
2. See for instance: http://www.extropy.org/, http://www.maxmore.com/extprn3.htm, http://humanityplus.org/learn/transhumanist-declaration/, accessed February 6, 2012.
3. For instance: *Le Courrier International* (n°806, April 13–29, 2006), *Le Figaro Magazine* (Saturday, January 8, 2011), *Time* (February 21, 2011), *Le Temps*

(n°4191, special issue, December 30 and 31, 2011, January 1 and 2, 2012), *Paris Match* (n°3271, January 26–February 1, 2012), *L'Illustré* (n°7, 2012).
4. See for instance the *Singularity University*, co-founded by Ray Kurzweil, one of the most prominent transhumanist defenders predicting the *singularity phenomenon*, which will bring humans into post-humanity (http://singularityu.org/, accessed January 11, 2012).
5. A. de Grey's claim in 'Rejuvenation biotech: can regenerative medicine defeat ageing?', *Slate*, posted on Monday November 15, 2010, http://www.slate.com/articles/technology/future_tense/2010/11/rejuvenation_biotech.html, accessed March 1, 2012.
6. I borrowed the term 'ideal of longevity' from Professor Samia Hurst of the Institute for Biomedical Ethics, University of Geneva. I also thank her for drawing my attention to the important difference between fighting death *per se* and fighting premature death.
7. *Deconstructing Harry*, 1997, written and directed by Woody Allen. Our emphasis.
8. http://www.ovid.com/site/products/ovidguide/medline.htm#geninfo, accessed February 15, 2012.
9. http://www.nlm.nih.gov/bsd/pmresources.html, accessed February 15, 2012.

References

Alexandre L. (2011) *La mort de la mort: comment la technomédecine va bouleverser l'humanité* (Paris: Jean-Claude Lattès).
Amundsen D.W. (1978) 'The physician's obligation to prolong life: a medical duty without classical roots', *The Hastings Center Report*, 8, 23–30.
Appel J.M. (2008) 'When the boss turns pusher: a proposal for employee protections in the age of cosmetic neurology', *Journal of Medical Ethics*, 34, 616–18.
Babcock Q. and Byrne T. (2000) 'Student perceptions of methylphenidate abuse at a public liberal arts college', *Journal of American College Health,*, 49, 143–5.
Barazzetti G. and Reichlin M. (2011) 'Life-extension: a biomedical goal? Scientific prospects, ethical concerns', *Swiss Medical Weekly*, 141, 1–7.
Becker E. (1973) *The Denial of Death* (New York: Free Press).
Benton J.P., Christopher A.N., Walter M.I. (2007) 'Death anxiety as a function of ageing anxiety', *Death Studies*, 31, 337–50.
Blais M. and Martineau S. (2006) 'L'analyse inductive générale: description d'une démarche visant à donner un sens à des données brutes', *Recherches Qualitatives*, 26, 1–18.
Blumenthal H.T. (2003) 'The ageing-disease dichotomy: true or false?', *The Journals of Gerontology*, 58, 138–45.
Bodner E. (2009) 'On the origins of ageism among older and younger adults', *International Psychogeriatric*, 21, 1003–14.
Bodner E., Bergman Y.S., Cohen-Fridel S. (2012) 'Different dimensions of ageist attitudes among men and women: a multigenerational perspective', *International Psychogeriatric*, 1–7.
Bolt L.L. (2007) 'True to oneself? Broad and narrow ideas on authenticity in the enhancement debate', *Theoretical Medicine and Bioethics*, 28, 285–300.
Bostrom N. (2005) 'In defense of posthuman dignity', *Bioethics*, 19, 202–14.

Boyer P. and Bergstrom B. (2011) 'Threat-detection in child development: an evolutionary perspective', *Neuroscience Biobehavioral Reviews*, 35, 1034–41.
Buchanan A. (2009) 'Human nature and enhancement', *Bioethics*, 23, 141–50.
Buyx A.M. (2008) 'Be careful what you wish for? Theoretical and ethical aspects of wish-fulfilling medicine', *Medicine Health Care Philosophy*, 11, 133–43.
Callahan D. (2000) 'Death and the research imperative', *The New England Journal of Medicine*, 342, 654–6.
Callahan D. (2002) 'How much medical progress can we afford? Equity and the cost of health care', *Journal of Molecular Biology*, 319, 885–90.
Canto-Sperber M. (2008) *Que peut l'éthique? Faire face à l'homme qui vient* (Paris: Textuel).
Caplan A.L. (2009) 'Good, better or best?', in Savulescu J. and Bostrom N. (eds) *Human Enhancement* (New York: Oxford University Press), pp. 199–209.
Chapman S. (2006) 'Erectile dysfunction and smoking: subverting tobacco industry images of masculine potency', *Tobacco Control*, 15, 73–4.
Crawford R. (2006) 'Health as a meaningful social practice', *Health* (London), 10, 401–20.
Daniels N. (2009) 'Can anyone really be talking about ethically modifying human nature?', in Savulescu J. and Bostrom N. (eds) *Human Enhancement* (New York: Oxford University Press), pp. 25–42.
de Grey A. (2005) 'A strategy for postponing ageing indefinitely', *Studies in Health Technology and Informatics*, 118, 209–19.
de Grey A. (2007) 'Life span extension research and public debate: societal considerations', *Studies in Ethics, Law, and Technology*, 1, 5. DOI: 10.2202/1941–6008.1011.
de Grey A. (2010) 'Hype and anti-hype in academic biogerontology research', *Rejuvenation Research*, 13, 137–8.
DeGrazia D. (2005) 'Enhancement technologies and human identity', *The Journal of Medicine and Philosophy*, 30, 261–83.
Drabiak-Syed, K. (2011) 'Physicians prescribing "medicine" for enhancement: why we should not and cannot overlook safety concerns', *American Journal of Bioethics*, 11, 17–19.
Ebrahim S., May M., Ben Shlomo Y., McCarron P., Frankel S., Yarnell J., Davey Smith G. (2002) 'Sexual intercourse and risk of ischaemic stroke and coronary heart disease: the Caerphilly study', *Journal of Epidemiology and Community Health*, 56, 99–102.
Elliott C. (2003) *Better than Well: American Medicine Meets the American Dream* (New York and London: W.W. Norton).
Elliott C. (2011) 'Enhancement technologies and the modern self', *The Journal of Medicine and Philosophy*, 36, 364–74.
Farrelly C. (2010) 'Why ageing research? The moral imperative to retard human ageing', *Annals of the New York Academy of Sciences*, 1197, 1–8.
Finch C. (2010) 'Secrets of a long life', *Nature*, 467, 274–5.
Fontana L., Partridge L., Longo V.D. (2010) 'Extending healthy life span: from yeast to humans', *Science* (New York), 328, 321–6.
Fries J.F. (1983) 'The compression of morbidity' (re-edn 2005), *Milbank Quarterly*, 83(4), 801–23.
Gallup G.G. Jr., Burch R.L., Platek S.M. (2002) 'Does semen have antidepressant properties?', *Archives of Sexual Behavior*, 31, 289–93.
Gaylin W. (1984) 'In defense of the dignity of being human', *The Hastings Center Report*, 14, 18–22.

Gaylin W. (1990) 'Fooling with mother nature', *The Hastings Center Report*, 20, 17-21.

Gems D. (2011) 'Tragedy and delight: the ethics of decelerated ageing', *Philosophical Transactions of the Royal Society of London: Series B, Biological Science*, 366, 108-12.

Greely H.T. (2005) 'Regulating human biological enhancements: questionable justifications and international complications', *UTS Law Review*, 7, 87 -110.

Greely H.T. (2012) 'Direct brain interventions to "treat" disfavored human behaviors: ethical and social issues', *Clinical Pharmacology and Therapeutics*, 91, 163-5.

Gruman G. (1966) 'A history of ideas about the prolongation of life: the evolution of prolongevity hypotheses to 1800', *Transactions of the American Philosophical Society*, 56, 1-102.

Gupta K. (2011) '"Screw health": representations of sex as a health-promoting activity in medical and popular literature', *Journal of Medical Humanities*, 32, 127-40.

Haas C.F., Champion A., Secor D. (2008) 'Motivating factors for seeking cosmetic surgery: a synthesis of the literature', *Plastic Surgical Nursing*, 28, 177-82.

Harris J. (2007) *Enhancing Evolution: The Ethical Case for Making Better People* (Princeton: Oxford and Princeton University Press).

Harris J. (2011) 'Moral enhancement and freedom', *Bioethics*, 25, 102-11.

Hayry M. (2011) 'Considerable life extension and three views on the meaning of life', *Cambridge Quarterly of Healthcare Ethics*, 20, 21-9.

Haseltine W. (2000) 'Futurology corner', *Science*, 290, 2249-49.

Helmuth L. (2005) 'Ageing research in the media. How the demands of newspaper and magazine publishing influence what people read about ageing', *EMBO Reports 6 Spec No*, S81-3.

Hotze T.D., Shah K., Anderson E.E., Wynia M.K. (2011) '"Doctor, would you prescribe a pill to help me...?" A national survey of physicians on using medicine for human enhancement', *American Journal of Bioethics*, 11, 3-13.

Juengst E.T., Binstock R.H., Mehlman M., Post S.G., Whitehouse P. (2003) 'Biogerontology, "anti-ageing medicine", and the challenges of human enhancement', *The Hastings Center Report*, 33, 21-30.

Kamm F. (2009) 'What is and is not wrong with enhancement?', in Savulescu J. and Bostrom N. (eds) *Human Enhancement* (New York: Oxford University Press), pp. 91-130.

Kaufman S.R., Mueller P.S., Ottenberg A.L., Koenig B.A. (2011) 'Ironic technology: old age and the implantable cardioverter defibrillator in US health care', *Social Science and Medicine*, 72, 6-14.

Kenyon C.J. (2010) 'The genetics of ageing', *Nature*, 464, 504-12.

Kiefer B. (2010) 'Médecine: la question des buts', *Revue Médicale Suisse*, 6 (266), 1936.

Kitcher P. (1997) *The Lives to Come: The Genetic Revolution and Human Possibilities* (London: Penguin).

Koch T. (2010) 'Enhancing who? Enhancing what? Ethics, bioethics, and transhumanism', *The Journal of Medicine and Philosophy*, 35, 685-99.

Kurzweil R. and Grossman T. (2009) 'Fantastic voyage: live long enough to live forever. The science behind radical life extension questions and answers', *Studies in Health Technology & Informatics*, 149, 187-94.

Ladle R.J., Malhado A.C., Todd P.A. (2007) 'Come all ye scientists, busy and exhausted. O come ye, O come ye, out of the lab', *Nature*, 450 (7173), 1156.

Lamkin M. (2011) 'Racist appearance standards and the enhancements that love them: Norman Daniels and skin-lightening cosmetics', *Bioethics*, 25, 185–91.

Lasher K.P. and Faulkender P.J. (1993) 'Measurement of ageing anxiety: development of the Anxiety about Ageing Scale', *International Journal of Ageing Human Development*, 37, 247–59.

Laure P. (2002) 'Les conduites dopantes: une prévention de l'échec?', *Psychotropes*, 8, 31–8.

Lehto R.H. and Stein, K.F. (2009) 'Death anxiety: an analysis of an evolving concept', *Research and Theory for Nursing Practice*, 23, 23–41.

Lentillon-Kaestner V. (2008) 'Conduites dopantes chez les jeunes cyclistes du milieu amateur au milieu professionnel', *Psychotropes*, 14, 41–57.

Lentillon-Kaestner V. and Brissonneau C. (2009) 'Appropriation progressive de la culture du dopage dans le cyclisme', *Déviance et Société*, 33, 519–41.

Lev O. (2011) 'Will biomedical enhancements undermine solidarity, responsibility, equality and autonomy?', *Bioethics*, 25, 177–84.

Liao S.M. (2011) 'Parental love pills: some ethical considerations', *Bioethics*, 25, 489–94.

Mahowald M.B. (2005) 'The President's Council on Bioethics 2002–2004: an overview', *Perspectives in Biology and Medicine*, 48, 159–71.

Mallia P. (2010) 'Clinical intervention in ageing: ethicolegal issues in assessing risk and benefit', *Clinical Interventions in Ageing*, 5, 373–80.

Martens A., Goldenberg J.L., Greenberg J. (2005) 'A terror management perspective on ageism', *Journal of Social Issues*, 6, 223–39.

Mauron A. (2005) 'The choosy reaper. From the myth of eternal youth to the reality of unequal death', *EMBO Reports 6 Spec No*, S67–71.

McKean E. (ed.) (2005) *The New Oxford American Dictionary*, 2nd edn (Oxford: Oxford University Press).

McMahon C.G. (2010) 'Get a better erection! – hope for sale – use sexual snake oil', *The Journal of Sexual Medicine*, 7, 1699–702.

Menuz V., Hurlimann T., Godard B. (2013) 'Is human enhancement also a personal matter?', *Science and Engineering Ethics*, 19, 161–77.

Moreno J.D. (2004) 'DARPA on your mind', *Cerebrum*, 6, 91–9.

Mykytyn C.E. (2006) 'Anti-ageing medicine: predictions, moral obligations, and biomedical intervention', *Anthropological Quarterly*, 79, 5–31.

Outram S.M. (2010) 'The use of methylphenidate among students: the future of enhancement?', *Journal of Medical Ethics*, 36, 198–202.

Pinkleton B.E., Austin E.W., Chen Y.C., Cohen M. (2012) 'The role of media literacy in shaping adolescents' understanding of and responses to sexual portrayals in mass media', *Journal of Health Communication*, 17, 460–76.

Powell R. and Buchanan A. (2011) 'Breaking evolution's chains: the prospect of deliberate genetic modification in humans', *The Journal of Medicine and Philosophy*, 36, 6–27.

President's Council on Bioethics (2003) *Beyond Therapy: Biotechnology and the Pursuit of Happiness* (Washington DC: Dana Press).

Pyszczynski T., Greenberg J., Solomon S. (1999) 'A dual-process model of defense against conscious and unconscious death-related thoughts: an extension of terror management theory', *Psychol Rev*, 106, 835–45.

Rae M.J., Butler R.N., Campisi J., de Grey A.D., Finch C.E., Gough M., Martin G.M., Vijg J., Perrott K.M., Logan B.J. (2010) 'The demographic and biomedical case for late-life interventions in ageing', *Science Translational Medicine*, 2, 1–6.

Raphael A. (2010) 'The ethics of cosmetic enhancement', *Pharos*, 73, 18–23.
Roco M.C. and Bainbridge W.S. (eds) (2003) *Converging Technologies for Improving Human Performance: Nanotechnology, Biotechnology, Information Technology and Cognitive Science* (Dordrecht: Kluwer Academic Publishers).
Rothman S.M. and Rothman D.J. (2004) *The Pursuit of Perfection: The Promise and Perils of Medical Enhancement* (New York: Vintage).
Safdar A., Bourgeois J.M., Ogborn D.I., Little J.P., Hettinga B.P., Akhtar M., Thompson J.E., Melov S., Mocellin N.J., Kujoth G.C., et al. (2011) 'Endurance exercise rescues progeroid ageing and induces systemic mitochondrial rejuvenation in mtDNA mutator mice', *Proceedings of the National Academy of Sciences of the United States of America*, 108, 4135–40.
Sahakian B. and Morein-Zamir S. (2007) 'Professor's little helper', *Nature*, 450, 1157–9.
Sandel M.J. (2004) *The Case against Perfection: Ethics in the Age of Genetic Engineering* (Cambridge MA: Harvard University Press).
Savulescu J. (2009) 'The human prejudice and the moral status of enhanced beings: what do we owe the gods?' in Savulescu J. and Bostrom N. (eds) *Human Enhancement* (New York: Oxford University Press), pp. 211–47.
Savulescu J. (2010) 'Human liberation: removing biological and psychological barriers to freedom', *Monash Bioethics Review*, 29, 01–18.
Savulescu J. and Kahane G. (2009) 'The moral obligation to create children with the best chance of the best life', *Bioethics*, 23, 274–90.
Schermer M. (2008a) 'Enhancements, easy shortcuts, and the richness of human activities', *Bioethics*, 22, 355–63.
Schermer M. (2008b) 'On the argument that enhancement is "cheating"', *Journal of Medical Ethics*, 34, 85–8.
Scheunemann L.P. and White D.B. (2011) 'The ethics and reality of rationing in medicine', *Chest*, 140, 1625–32.
Schneider A.J. (2006) 'Cultural nuances: doping, cycling and the tour de France', *Sport in Society*, 9, 212–26.
Sparrow R. (2011) 'A not-so-new eugenics: Harris and Savulescu on human enhancement', *Hastings Center Report*, 41, 32–42.
Spinney L. (2006) 'Gerontology: eat your cake and have it', *Nature*, 441, 807–9.
Turner L. (2004) 'Biotechnology, bioethics and anti-ageing interventions', *Trends in Biotechnology*, 22, 219–21.
Van Hilvoorde I. and Landeweerd L. (2010) 'Enhancing disabilities: transhumanism under the veil of inclusion?', *Disability and Rehabilitation*, 32, 2222–7.
Vijg J. and Campisi J. (2008) 'Puzzles, promises and a cure for ageing', *Nature*, 454, 1065–71.
Warren O.J., Leff D.R., Athanasiou T., Kennard C., Darzi A. (2009) 'The neurocognitive enhancement of surgeons: an ethical perspective', *Journal of Surgical Research*, 152, 167–72.
Wijsbek H. (2000) 'The pursuit of beauty: the enforcement of aesthetics or a freely adopted lifestyle?', *Journal of Medical Ethics*, 26, 454–8.
Wilmoth J.R. (2000) 'Demography of longevity: past, present, and future trends', *Exp Gerontol*, 35, 1111–29.
Wimo A. and Prince M. (2010) *World Alzheimer Report: The Global Economic Impact of Dementia* (London: Alzheimer's Disease International).

4
The Moral Ambiguity of Human Enhancement
Ruud ter Meulen

Biotechnologies are generally developed to heal people from severe diseases. However, many of these technologies have the potential to be used beyond the frame of therapy as a way to improve or *enhance* normal human capacities. Biotechnologies can help to make people think better, to improve their memory and perception, to feel happier or to improve physical skills in sports, music or dance, or to extend the normal human lifespan. In view of the potential of biotechnologies (and other technologies like nanotechnologies and information technologies) to change our capacities, there is an ethical debate whether such an enhancement may alter our sense of self, our human nature and our relation with other life forms. Moreover, there is a concern about the impact of these technologies on our society and the position of vulnerable groups. More fundamentally there is concern over whether enhancement is a good thing in itself and whether it may expose our human nature, our personal life and our society to irreversible damage.

The moral ambiguity of enhancement

Although enhancement basically means improvement or adding of a new capacity, its application in the area of human cognitive and physical capacities seems to evoke negative sentiments to many. Enhancement is an ambiguous concept which can mean better and more, but also something that many people may think to be less desirable and that should be avoided. For example, genetic doping in sports may be highly valued by some athletes, but may be judged negatively by many others because it may result in unfair competition. Indeed, there is some concern that the use of cognitive and physical enhancers reinforces individualist values, such as competitiveness and cheating behaviour. There is also a concern that individuals may feel forced to take enhancing drugs to keep up with this competition and that this is problematic as it promotes a high-risk, coercive culture.

The negative evaluation of enhancement is partly the result of the history of eugenics in the early decades of the 20th century. Many handicapped persons and people with psychiatric and psychogeriatric disorders were sterilised and killed because they did not meet the Nazis' ideals of race and humanity. People with physical and learning disabilities have the feeling that the application of enhancement technologies as a new kind of eugenics might result in a less favourable view of those with handicaps and disabilities, and even of their existence as members of society. One can argue that some kinds of enhancement can indeed be considered as eugenic. An example may be the genetic improvement and genetic selection through prenatal and pre-implantation genetic diagnosis (PGD and PIGD). However, the proponents of these new technologies argue that there is a big difference with the approach of the 'old' eugenics because of the emphasis in the 'new' eugenics on free choice and autonomy ('liberal eugenics', Agar, 2004). Nonetheless, the basic idea is the same, namely the weeding out of undesirable physical and psychological traits. One should be aware that a broad application of enhancement can indeed lead to discrimination of handicapped persons.

But is enhancement always bad or at least dubious and thus unacceptable? To reach a better understanding of the moral value of enhancement, its goals are often compared with those of therapy. While therapeutic interventions are usually the source for the development of enhancement technologies, our moral evaluation of each differs markedly. Where therapy is generally seen as something good, perhaps for its relief of suffering, enhancement is sometimes considered bad, or at least ambiguous. In this debate an abstract conceptual scale is often invoked with, at one end, examples of enhancement and, at the other end, the normal applications of medical technology. However, there is no consensus about what should be seen as a 'normal' application of medical technology and this is one of the major difficulties with the rejection of human enhancement. A normal application is defined as a treatment that falls within the goals of medicine, like the treatment of disease and the alleviation of suffering. Yet establishing robust distinctions is not so easy. Take, for example, the following case, outlined by Norman Daniels in a classic essay:

> Johnny is a short eleven year old boy with documented growth-hormone deficiency resulting from a brain tumour. His parents are of

average height. His predicted adult height without growth hormone (GH) treatment is approximately 160 cm (5 feet 3 inches). Billy is a short eleven year old boy with normal GH secretion according to current testing methods. However, his parents are extremely short, and he has a predicted adult height of 160 cm (5 feet 3 inches). (Daniels, 1992, p. 46)

Johnny's shortness is a function of his disease, while Billy has a normal genotype, one that produces normal levels of GH. However, both boys are similar as they are both short and suffer from this shortness, as our society values tall stature. In Johnny's case, one could say GH treatment falls within the goals of medicine and is an acceptable treatment. However, for Billy there is no disease. In his case GH treatment can be considered an enhancement, and thus a treatment that falls outside the medical domain (Parens, 1998).

Moreover, therapy and enhancement, to a certain extent, overlap; all successful therapies are a kind of enhancement, even if not all enhancements can be called therapeutic. The improvement and regeneration of organs and tissues in the elderly may be seen as enhancement, but it can be considered to be a therapy as well. The therapy–enhancement distinction is also complicated because of social, cultural and historical differences in the understanding of health, disease and disability. Engelhardt (1974), for example, argues that concepts of health and disease are guided by value judgements and prejudices that may change over time. Examples are masturbation which was for a long time considered a disease, and homosexuality which was only recently removed from the Diagnostic and Statistical Manual of Psychiatric Disorders (DSM) of the American Psychiatric Association (APA). Efforts to limit access to certain technologies because they fall outside the prevailing definitions of disease can be considered to be biased and an inconsistent application of our values. This could imply that the distinction between normal and enhanced is basically normative rather than objective or universally valid. One can also argue that the distinction between treatment and enhancement is based on a cultural definition of what counts as a disease and what does not.

Utilitarian authors, like for example John Harris (2007), argue that enhancement needs to be evaluated in consequentialist terms, in the same way as with therapeutic interventions. Both therapy and enhancement should be evaluated in terms of benefits and harms and, in this respect, no distinction should be made between therapy and enhancement. Harris has a rather positive view on enhancement as he believes that the development and use of human enhancement technologies will improve our human situation. As such, human enhancement is no different to the way humans have tried to improve their situation in the past centuries. According to Harris we have a moral duty to enhance ourselves as we have a duty to make things better for ourselves and others.

However, and this is a problem for utilitarianism in general, it is not clear what count as 'benefits' and what as 'harms'. Though this is already difficult in the case of therapy, it is even more problematic in the case of enhancement. This concept is strongly linked to individual views on personal identity and to personal preferences for specific lifestyles. In the case of mood enhancement, for example by the use of anti-depressants, it is not clear whether having only cheerful and happy moods is a benefit without any harm, though one could imagine how insufferable a permanently cheerful person could be. Emotions can have a positive and a negative value, but both 'positive' and 'negative' emotions have a function in our individual and social life. Supposedly 'bad' feelings such as the emergency emotions of fear and rage are important for survival, coping and adaptation. Other bad feelings, like guilt and shame, serve conscience and communal life in various positive ways. Good feelings such as joy and pride serve aspiration, achievement and quality of life. Thus, so-called good and bad feelings, emotions or mood states are important for alerting, liberating and enriching individual and communal life. On this basis, it is difficult to claim, *a priori*, what kind of mood state should be picked out 'to be enhanced' (Berghmans et al., 2011). Indeed, it is sometimes only in retrospect that we can know whether a particular emotional experience was generally positive or negative and only then in the context of the act of memory (which we might also be tempted to alter).

Enhancement and human nature

Apart from the evocation of the eugenic movement, enhancement is considered morally problematic by some because it challenges the concept of human nature: for them the distinction between enhancement and therapy is essential. Therapy is considered natural and enhancement unnatural; therapy is seen as a way to assist the natural healing process, whereas enhancement is adding something – possibly detrimental – to the human being that is considered unnatural (President's Council on Bioethics, 2003). However, proponents of enhancement also justify their arguments on the basis of the concept of nature: they claim that the desire and pursuit of enhancement by the development of new technologies is part of human nature and must be considered as morally good. Given man's nature, there is not only a capability but a moral obligation to enhance one's capacities.

In order to understand why some authors state that enhancement of human capacities is a violation of human integrity and others see it as a moral obligation, it is necessary to look in detail at the perceptions of human nature that underlie the debate. According to the proponents of the moral distinction between enhancement and therapy, the ethically suspect element of 'enhancement' lies within the idea of changing human nature through the addition of non-native skills or qualities. When a physician intervenes therapeutically to correct some deficiency or deviation from a patient's natural wholeness, he acts as a servant to the goal of health and as an assistant to nature's own powers of self-healing, themselves wondrous products of evolutionary selection. But when a bioengineer intervenes for non-therapeutic ends, they stand not as nature's servant but as aspiring master, guided by nothing but their own will and serving ends of their own devising (President's Council on Bioethics, 2003).

In this view, nature is seen as a process (that is the reference to self-healing and evolutionary selection) that constitutes the human being as well as the world as a whole. This approach is reminiscent of the Aristotelian definition of nature as a self-moving process. In his lecture on *Physics*, Aristotle explains nature as an ongoing process mastered by an intrinsic causality. In this approach, the human being is part of and dependent on cosmic natural process and can only to a limited extent master their own physical nature.

However, in a different perspective on enhancement, some authors make use of a distinction between nature and human nature. They argue that, as opposed to other living beings, only human beings possess the potential to master their physical and cognitive capacities to an extent that goes beyond the element of natural chance. The idea of the human being as capable of mastering nature dates back to the Enlightenment and the Cartesian philosophical tradition. The ability to act according to free will is what distinguishes the human being from other living beings. At the same time it is exactly this ability that constitutes the moral dilemma, as free will makes it possible for human beings to act against the laws or processes of nature. But is intervening in the natural process for non-therapeutic ends morally wrong because of an intrinsic value of the natural process itself? Or is it rather the element of 'aspiring master' that causes ethical concern?

Enhancement can be interpreted both as part of human nature and as a potential violation of human nature. Some would claim that nature and natural process are fragile, and that any interference in them must be approached with precaution. The unknown long-term effects of the new enhancement technologies make our own desire for enhancement of human capacities a potential danger for present and future generations. Others would claim that the desire for enhancement of human capacities is to be pursued: in the end, this search for ways to enhance ourselves is a natural part of being human. The urge to transform ourselves has been a force in history as far back as we can see. It's been selected for by millions of years of evolution. It is wired deep in our genes – a natural outgrowth of our human intelligence, curiosity and drive. To turn our backs on this power would be to turn our backs on our true nature (Naam, 2005).

A radical elaboration of this approach is represented in the idea of *transhumanism* or *post-humanity*. Proponents of this line of thought claim that enhancement of existing human capacities will make us overcome our current vulnerability, and will result in a post-human state which is no longer dependent on our natural surroundings (Bostrom, 2005a). The possibility, for example, of uploading our brains like computers and growing new organs by using human stem cells will eventually make us able to deal with human vulnerability to disease and damage. However, such radical alterations will change the core nature of *Homo sapiens* and turn us into post-humans.

However, such a development may create new divisions in our society. This view is taken by Francis Fukuyama in his book *Our Posthuman Future* (2002). He is worried about the emergence of new genetic classes, which could lead to increasing inequality in our society. Fukuyama thinks that genetic enhancement may lead to a more egalitarian society, particularly when access to these technologies is funded for everybody, but there is also a serious danger that it will divide our society into different genetic classes. Genetic enhancement will 'not be a threat to the dignity of normal adult human beings but rather to those we have defined as characterising human specificity. The largest group of beings in this category are the unborn, but it could also include infants, the terminally sick, elderly people with debilitating diseases, and the disabled' (Fukuyama, 2004, p. 174). Genetic enhancements then will put increasing emphasis on intelligence, cognitive capacities and sensitive emotions as defining characteristics for dignity and humanity. Humans that do not have these (enhanced) capacities will be seen as inferior and as possessing fewer human rights.

Fukuyama's analysis raises serious concerns about the quality of our future relationships in the 'posthuman society'. If genetic enhancement and other kinds of enhancement are not integrated in a broader moral worldview on human personhood and dignity, they might indeed result in an increasing individualisation of our society and a decrease of solidarity with the weaker and vulnerable groups. A similar concern can be found in the book *The Case against Perfection* by Michael Sandel (2007). Sandel argues that we have become *too* responsible for our own fate. 'Parents have become responsible for choosing, or failing to choose, the right traits for their children. Athletes become responsible for acquiring, or failing to acquire, the talents that will help their team win' (Sandel, 2007). This 'explosion of responsibility' has come at the expense of an attitude of acceptance of our limitations and of the notion of 'giftedness'. Sandel argues that there is a connection between the notions of 'giftedness' and 'solidarity': as soon as we are aware of the contingency of our gifts, we will develop our capacity of seeing ourselves as sharing a common fate. Bioengineering our children and ourselves is not just a matter of autonomy, individual rights and power: 'changing our nature to fit the world, rather than the other way around, is actually the deepest form of disempowerment. It distracts us from reflecting critically on the world, and deadens the impulse to social and political improvement' (Sandel, 2007). In a comparable way, Habermas (2003) argues that the use of pre-implementation diagnosis to select an embryo according to the wishes of the parents is an instrumentalisation of the creation of human life and an intention to master the contingency (or 'giftedness' in Sandel's words) of human nature by a third party (the parents). Such intergenerational control is a serious breach of the self-understanding of the individual that has come into existence by way of this procedure.

For transhumanists, the human body is not part of our fate (or what is given to us) but is 'brute matter' which we can manipulate and change without any impact on our self-understanding (Edgar, 2009). Habermas, Sandel and Fukuyama argue instead that intervening in the body for the purpose of enhancement is not a morally neutral procedure: it is an attempt to control the contingency of our nature with important consequences for the self-understanding of the individual and for interpersonal relations in our society.

Enhancement and authenticity

In his book *Better than Well*, Carl Elliott tries to answer the question of why enhancement technologies have become so popular in contemporary society (Elliott, 2003). According to Elliott, the answer can be found in the ideal of authenticity which has become so dominant in our society. Elliott refers to the work of Charles Taylor, who tried to uncover the roots of this ideal in the history of western culture (Taylor, 1991). According to Taylor, this ideal meant that to reach an understanding of our moral directions and obligations we should have contact with our feelings. Originally, this ideal was formulated by Romantic philosophers in the early 18th century. Particularly important is the work of the German philosopher Herder, who put forward the idea that every person has an original way of being human and that this difference has an important moral significance. Taylor summarises this view as follows: 'There is a certain way of being human and that is my way. I am called to live my life in this way, and not in imitation of anyone else's. But this gives a new importance to being true to myself. If I am not, I miss the point of my life, I miss what is being human for me' (Taylor, 1991). Authenticity has become a powerful ideal in our culture; it means that we have to get in contact with ourselves, with our own inner nature, our inner voice, particularly when there is a threat it will become lost because of the pressure of external conformity. It is strongly linked to the idea of originality; each voice has something of its own to say. External sources of moral conduct are not important: I can only find these directions within myself.

According to Elliott, it is the ideal of authenticity that drives the language of people using biotechnologies and other technologies to enhance their functioning. From Prozac to face-lifts, using these technologies or drugs make people feel more authentic, more 'being themselves' (Elliott, 2003). He gives the example of Jan Morris, who transformed from a man into a woman. She felt after the sex-reassignment surgery that she had found her true self and that her life became fulfilled by the procedure. She was very happy because of this, and 'fulfilment requires being true to that inner voice' (Elliott, 2003). Elliott also gives the example of persons who had their limbs amputated at their own request: 'I have always felt I should be an amputee. It is a desire to see myself, be myself, as I "know", or "feel" myself to be' (Elliott, 2003).

A critique of the preoccupation with the self is voiced by communitarian philosophers who argue that the self can only realise itself in the community with others. In contemporary culture, the search for the self has resulted in narcissism and estrangement from others, a process that is reinforced by the use of enhancement technologies. In its discussion of the role of Prozac and other Selective Serotonin Reuptake Inhibitors (SSRIs) in mood enhancing and mood elevation, the 2003 US President's

Council states that, with the availability of mood-enhancing drugs, individuals become so preoccupied with their state of mind 'that they remove themselves increasingly from active participation in civic life, discarding those attachments without which they cannot achieve the happiness they seek and without which the community cannot survive and flourish' (President's Council on Bioethics, 2003). The danger of a widespread use of mood enhancing drugs, according to the Council, involves the 'solipsistic self worried only about the state of his feelings, who uses psychopharmacology to ensure a flat and shallow self regarding psychic pleasure'.

Taylor agrees that the ideal of authenticity in its modern variant has indeed been transformed into an inward-looking, solipsistic kind of individuality. According to Taylor, this decline into solipsism must be seen as a deviation from and as a wrong interpretation of the original idea of authenticity. Originally, the change to authenticity was considered a more subtle and personal way of connecting to a larger whole, be it society or nature or the cosmos. It was a new way or manner of relating to the social and (supra-) natural world. This meaning of authenticity has gradually been hidden and forgotten, in favour of an obsession with the manner itself (instead of the goal). Real authenticity is reached only by a relation to a world outside of us, or, in Taylor's words, a *horizon of meaning* that exists independent of us. Taylor makes a distinction between 'boosters' and 'knockers' of the ideal of authenticity. 'Boosters' defend the ideal of 'finding yourself'. They focus on subjectivism, moral relativism and non-interference with personal life by others. 'Knockers' criticise the ideal, as it would lead to individualism, atomism, denial of citizenship, duties of solidarity and the needs of the natural environment. This is the view of communitarians and the President's Council on Bioethics. According to Taylor, both 'boosters' and 'knockers' have got it wrong, or are at least one-sided in their interpretation of the ideal of authenticity. The boosters have got it wrong, because they focus on the individual and the self, without considering the link with the outside world. The knockers make the mistake of denying the importance of individual development and authenticity as a genuine and important cultural ideal. The enhancement users can be considered 'boosters': they look for their inner, real self but forget about the world outside them. The knockers can be found in the President's Council for Bioethics and among communitarians, who reject the efforts towards enhancement as a social evil.

However, the President's Council on Bioethics might be too pessimistic about the way enhancement reduces social and civic life. It is debatable whether the use of anti-depressants leads to 'affective blunting' (Furlan et al., 2004). In fact, the use of these drugs can even help people to relate better with their spiritual and social world. In his contribution to *Prozac as a way of life*, Tod Chambers gives examples of how spiritual lives have been changed by the impact of Prozac. After taking Prozac, one devout Christian felt 'like living again. And I began to experience God like I never had before' (Chambers, 2004). In his book *Listening to Prozac*, Peter Kramer (1994) gives examples of how patients by using Prozac felt more able to cope with ordinary life, overcoming shyness and social inhibition.

Enhancement technologies do not necessarily create solipsistic, selfish individuals, searching for their own selves. If there is a tendency of individuals to use enhancement technologies for this purpose, it is not the technology that is to blame, but the social and cultural process that is driving individuals into this sort of behaviour. Enhancement can very well be applied to 'open up' individuals to play an active life in the community and to search for meaning by relating to moral and spiritual sources.

Enhancement and human dignity

One of the key criticisms raised against enhancement is that it may hinder the development of individuals into moral agents who are capable of developing their own set of values and of entering into meaningful social relationships. According to this view, enhancement technologies may come to dominate people's lives replacing individual agency and responsibility by a slavish use of drugs and technologies to enhance their capacities. An example is the critical paper by Darren Shickle (2000) who argues that genetic enhancement technologies may 'bypass' the struggle involved in many activities. Following the ideas of Joseph Amato about the value of suffering, Shickle argues that struggle is involved in many activities and that human activities are valued in relation with the amount of suffering involved: 'Struggle is a measure of human activity, and even pleasurable activities require efforts that result in pain [...] The more that we have to struggle to attain a thing, the stronger the claim to possession. It is the process of striving that makes things worthwhile' (Shickle, 2000). In other words, the use of enhancement technologies results in eliminating struggle in our existence, and by consequence an inability to develop a sense of value and of meaningful relationships with others.

Similarly Fukuyama (2003) argues that the reduction of human qualities for utilitarian purposes will affect our dignity, which he defines as a range of human qualities and abilities that connect us to other human beings. The utilitarian goal of minimising suffering leads to a destruction 'of the way we react to, confront, overcome, and frequently succumb to pain, suffering and death' (Fukuyama 2003). In the absence of these human evils, we lose qualities like sympathy, compassion, courage, heroism, solidarity or strength of character which are essential for human dignity.

These comments remind us of the analysis by the German sociologist Max Weber, and later on by the *Frankfurter Schule*, of the role of technology and of rationalisation in modern society. According to Weber, our social life is increasingly dominated by the laws of instrumental reason, which means the emphasis on calculation, prediction, effectiveness, bureaucracy and control as the basic principles of social life. According to Weber, instrumental reason has resulted in a greater control and improvement of our economic and social circumstances. However, it has also resulted in a disappearing of the sense of meaning in our natural and social world ('Entzauberung der Welt'). We are deprived of the capacities to experience meaning in the world, but also to share moral values with each other. But does this process necessarily lead to a diminishing of human dignity? Bostrom (2005b; 2008) argues that, on the contrary, the increasing improvement of our human capacities will lead to a further increase of our dignity, in the sense of mastery of our existence and of the world around us.

Perhaps it makes sense to get a better understanding of the concept of human dignity which is so often invoked in ethical debates but which lacks a precise definition.[1] Taylor analyses the meaning of dignity by going back to its original use in modern society. In pre-modern times, the role and importance of individuals in society were defined by the concept of 'honour', a characteristic or reward that was not open to all members of the ancient regime (Taylor, 1991). While the concept of honour was intrinsically linked to inequalities, the concept of dignity was the only concept that was compatible with democratic societies in which everybody had the chance to be recognised as a person with specific value. Against the notion of honour in the old order, the concept of dignity in democratic societies is according to Taylor essentially egalitarian and universalist. The change to dignity as the new way of social recognition has had important repercussions for the way individuals define their identity: identity in modern times is not socially derived from a pre-existing order as was the case in the *Ancien Régime*. Identity in modern times is generated and socially recognised in a process of dialogue and negotiation.

Dignity in modern society means that one's identity is recognised by significant others, including the differences one has developed compared to other individuals. This policy of recognition means also that every individual, no matter how different its identity, is treated with equal value. Modern accounts of dignity acknowledge the connection between dignity and equality, but they generally ignore the moral background we need for doing so. According to Taylor, to come to a mutual recognition of the equal value of different identities requires that we share some standards of value on which the identities concerned check-out as equal. Recognising individuality and dignity requires more than the fair and equal treatment of individuals, for example in (libertarian) concepts of justice: it requires a joint project in which individuals express and rank their values in a shared 'horizon of significance' in order to recognise difference, equality and dignity.

On the basis of this understanding of dignity, there is no reason why the use of technology or enhancement-technology should obstruct the recognition of dignity. For example, technological devices like electronic wheelchairs, human–computer interaction tools, and hearing and visual aids, are enabling people with disabilities to participate in society and to be recognised as persons with dignity. Technology can help to promote individuality and human relationships, as long as they are informed by what Taylor calls 'the ethic of practical benevolence', meaning the affirmation of daily production and reproduction (Taylor, 1991). Practical benevolence is morally significant as it helps to establish a human community with meaningful social relationships.

However, there is an inherent tendency in technology to dominate and control our lives and to fragmentise and de-humanise society, a process which Max Weber called the 'iron cage of technology' (Weber, 1978). Technology can indeed take over human attributes and may promote standardisation, destroying what is unique in each individual. An enhanced human being then is at risk of being so closely controlled by technological devices that this being becomes stripped of any individuality. However, such de-individualisation is not a necessary consequence of applying technology, not even of enhancement technologies. As long as we integrate technology into our lives as a benevolent force, it may help to promote dialogue, individuality and recognition of dignity. Enhanced human beings should be able to connect with other human beings and should be able to integrate technology in their own narrative understanding of themselves. They also should be able to develop human relationships which are essential for the recognition of their individuality and dignity. I can even imagine that enhancement technologies can promote and *improve* dignity by supporting dialogue and human understanding.

Conclusion

The goals of enhancement technologies are generally formulated in utilitarian terms like diminishing of suffering and greater and more effective control of our contingency. But *why* we are trying to achieve this control and *to what extent* and *to which goal* is not formulated within the enhancement technologies themselves. Such an understanding can only be achieved by discussing the contribution these technologies might make to the improvement of our own lives, our personal relationships and our society. Enhancement technologies need not necessarily result in loss of dignity and authenticity, but can open up individuals and can improve social relationships if they are placed within a shared perspective of values or a horizon of meaning. For this reason, the development and use of these technologies should not be left to the free forces of our society. The introduction of these technologies should be guided by a shared understanding of how these technologies might affect fundamental values like solidarity, dignity and responsibility. To reach such an understanding will be important *before* such technologies have become overly fashionable and irreversible harms have occurred to individuals and our society.

Notes

This chapter is an original, amended English version of "Sullo Human Enhancement" published in Luca Grion (Ed.), *La sfida postumanista. Colloqui sul significato della tecnica*, Bologna, Il Mulino, 2012, pp. 129-49. I thank the Centro Interdipartimentale di Ricerca e Servizi per le Decisioni Giuridico Ambientali e la Certificazione Etica di Impresa (CIGA), the editor of the volume, and Il Mulino, the publisher, for their permission to publish this English version.

1. The following section was previously produced in: Meulen, R. ter, (2010) 'Dignity, posthumanism and the community of values. Answer to Fabrice Jotterand and Nick Bostrom'. *American Journal of Bioethics*, 10(7), 69-70.

References

Agar N. (2004) *Liberal Eugenics. In Defence of Human Enhancement* (Oxford: Blackwell).
Berghmans R., ter Meulen R., Malizia A., Vos R. (2011) 'Scientific, ethical, and social issues in mood enhancement' in Savulescu J., ter Meulen R., Kahane G. (eds) *Enhancing Human Capacities* (Oxford: Wiley-Blackwell), pp. 153-65.

Bostrom N. (2005a) 'A history of transhumanist thought', *Journal of Evolution and Technology*, 14(1), 1–27.
Bostrom N. (2005b) 'In defense of posthuman dignity', *Bioethics*, 19, 202–14.
Bostrom N. (2008) 'Dignity and enhancement' in *Human Dignity and Bioethics. Essays Commissioned by the President's Council on Bioethics* (Washington DC), pp. 173–206.
Chambers T. (2004) 'Prozac for the sick soul' in Elliott C. and Chambers T. *Prozac as a Way of Life* (Chapel Hill: University of North Carolina Press).
Daniels N. (1992) 'Growth hormone therapy for short stature', *Growth: Genetics and Hormones*, 8 (supplement), 46–8.
Edgar A. (2009) 'The hermeneutic challenge of genetic engineering', *Medicine, Health Care and Philosophy*, 12, 157–67.
Elliott C. (2003) *Better than Well. American Medicine Meets the American Dream* (New York and London: Norton & Co.).
Engelhardt H.T. (1974) 'The disease of masturbation: values and the concept of disease', *Bulletin of the History of Medicine*, 48, 234–48.
Fukuyama F. (2003) *Our Posthuman Future. Consequences of the Biotechnology Revolution* (London: Profile Books).
Furlan P.M., Kallan M.J., Have T.T. et al. (2004) 'SSRIs do not cause affective blunting in healthy elderly volunteers', *American Journal of Geriatric Psychiatry*, 12(3), 323–30.
Habermas J. (2003) *The Future of Human Nature* (Cambridge: Polity Press). (Translation of Habermas, J. (2001) *Die Zukunft der menschlichen Natur. Auf dem Weg zu einer liberalen Eugenik?* Frankfurt am Main: Suhrkamp Verlag)
Harris J. (2007) *Enhancing Evolution* (Princeton: Princeton University Press).
Kramer P. (1994) *Listening to Prozac* (London: Fourth Estate).
Naam R. (2005) *More than Human. Embracing the Promise of Biological Enhancement* (New York: Broadway Books).
Parens E. (ed.) (1998) *Enhancing Human Traits. Ethical and Legal Implications* (Washington: Georgetown University Press).
[US] President's Council on Bioethics (2003) *Beyond Therapy – Biotechnology and the Pursuit of Happiness, a Report of the President's Council on Bioethics* (Washington DC: US Government Office), www.bioethics.gov.
Sandel M. (2007) *The Case against Perfection. Ethics in the Age of Genetic Engineering* (Cambridge MA: Belknap Press of Harvard University Press).
Shickle D. (2000) 'Are "genetic enhancements" really enhancements?', *Cambridge Health Ethics Quarterly*, 9, 342–52.
Taylor C. (1991) *The Ethics of Authenticity* (Cambridge MA: Harvard University Press).
Weber M. (1978) *Economics and Society: An Outline of Interpretive Sociology* (Berkeley: University of California Press).

Part II

Learning from Enhancement Practices

5
A Scale and a Paradigmatic Framework for Human Enhancement

Pascal Nouvel

Introduction

I would like to propose a scale for enhancement techniques, and in so doing, additionally show that a set of techniques, already in use or likely to be used soon, can serve as a paradigmatic framework for such a scale. As a point of departure, I will briefly comment on a generally accepted definition of 'human enhancement': enhancement refers to any attempt to temporarily or permanently overcome the current limitations of the human body through natural or artificial means. This is a comprehensive definition that would include both an apparatus designed to improve the vision of persons suffering from nearsightedness and the engineering of the gene coding for myostatin in a human embryo so as to give birth to an engineered champion.[1] Two points in the definition may require further attention: the concept of limitation and the distinction between natural and artificial.

It is clear that a definition containing a statement such as 'overcome the current limitations' also implicitly contains the idea that the human body is intrinsically limited. But limited compared to what? What performance can be taken as a standard with respect to which a 'limitation' can be defined? One can certainly consider a human body affected by some illness (or injury) as limited because one can compare the performance of this body with the performance it exhibited before the person was ill (or injured). Indeed a point of comparison seems to be necessary to attribute a 'limitation' to any living body. We do not say that a plant is limited because it cannot move. But we do say so when an illness has deprived a human being of the possibility of moving autonomously.

As Georges Canguilhem (1989) suggested in his influential book *The Normal and the Pathological*, a 'feeling of limitation' might be sufficient to define an actual limitation. If I want to climb to the top of a mountain and I stop my ascent a few hundred metres before the summit because my body can no longer move ahead, I experience a limit with respect to what my body can actually do. I also 'feel' a limit which comes from the comparison between what I desire and what my body can actually do. Such a feeling of limitation appears when, experiencing an illness, a person becomes aware that things he or she was used to doing without difficulty have become painful or even impossible to do. Here, the feeling of limitation comes from a discrepancy between the performance a person was used to eliciting from his or her body and those that are possible later with that same body. But when one speaks about a 'normal' body, it is conventionally supposed that such a discrepancy does not exist (otherwise, one would not say that it is 'normal'). In other words, this discrepancy is precisely what is called an 'illness'.[2] So, what kind of illness is associated with the 'limitation' evoked in the definition of enhancement? Can the 'feeling of limitation' be produced, for instance, by the social environment in such a way that a normal organism could suffer from its 'normality'? What kind of tool could then be used to overcome the limitation given by one's own 'normality'? Among those tools, is it possible to distinguish between those that are natural and those that are artificial? More generally, how relevant are these notions to the debate on human enhancement?[3]

The distinction between natural and artificial

At first glance, the distinction between a natural and an artificial object is trivial: a tree in the countryside is a natural object, as is the bird singing on one of its branches. On the contrary, the bicycle that has been left by someone against the trunk of the tree is artificial: it could not exist if a human had not built it. Thus, as Aristotle put it a long time ago, the world can be divided into natural and artificial things and the distinction (i.e. the category in which one should put any given object) is not problematic: for any object we can, according to Aristotle, find quickly and unequivocally to which category, natural or artificial, it belongs (Aristotle, 4th century BC, Book 1, 192b, p.32 ff.).

However, some philosophers have recently questioned this distinction in the light of what modern science allows us to know and to do with things. For instance, Bernadette Bensaude-Vincent and William

Newmann, in a recent book entitled *The Natural and the Artificial* (2007), write: 'with each passing day the traditional boundary between the natural and the artificial becomes less distinct'. It might thus not be only by chance that some contemporary artists are mixing together the natural and the artificial in order to obtain sometimes astonishing effects, as a result, at least in part, from the crossing of boundaries between those two fundamental categories. Examples of such a combination of natural and artificial objects in modern art are easy to find: see, for instance, the shark in a transparent Plexiglas box by Damien Hirst (Gether and Hirst, 2009), where a natural being – the shark – together with an artificial one – the transparent box – form a piece of art.

In the contemporary debate on enhancement, the distinction appears to have a very variable relevance depending on the type of enhancement considered, as one will see later. The distinction between natural and artificial, although it is arguably blurred in certain instances, might nevertheless appear useful to more clearly define the term 'enhancement' itself. Indeed, the definition of the term covers such a diversity of possible interventions that one might ask if there is something common to all the 'natural or artificial' means that are encompassed in this single concept. One of the recurring problems with enhancement admittedly comes from the fact that one speaks about many different things through a single and poorly determined concept. What, precisely, falls under the concept of enhancement? Which techniques? How can these techniques be characterised?[4]

In order to address these questions, it is useful to examine whether it is possible to put into some order the various possible actions that one can carry out on the human body. Moreover, it appears helpful to define a 'degree of artificiality' of the techniques that are used to modify the human body's performances. A paradigmatic example of an enhancing technique (i.e. a single and well-documented example that covers all possible aspects of the problem) would also be of great value. Accordingly, we will first try to establish a reasoned catalogue of enhancement techniques.[5] We will then try to find a single representative illustration of it (a paradigmatic example).

A catalogue of enhancement techniques

If one looks at techniques that could 'overcome the limitations of the human body', one will notice that it always consists of the introduction of a link between some object and the human body.[6] The object can have different sizes and shapes, as well as different actions when it is in or near the human body. It can vary from a prosthesis to a gene. But there is always an object that is introduced into the body or attached to it (I am thus not considering here the techniques designed to produce social enhancement of health, such as the enhancement obtained through hygienic methods in health care at the end of the 19th century).

Amongst the objects attached to or introduced into the body for the sake of enhancing some of its performance, one can first distinguish those that remain located in a certain region of the body (this is the case of any implant or prosthesis) and those that dissolve in the body (this is the case of any absorbed or injected substance). Accordingly, the degree of localisation of the object introduced into the body can provide a first criterion that can help to organise the various modes of 'enhancing' techniques. The apparatus that Oscar Pistorius uses as legs, for instance, are *localised*. A doping drug, such as an amphetamine, is *de-localised*: after having been absorbed, it will be present in the whole body for a while.

I wrote 'for a while': this expression introduces a second criterion that can be used to organise performance-enhancing substances or methods, which focuses on the duration of the enhancement effects produced. Indeed, the enhancement effect can be either temporary or definitive. When the marathon runner Emile Zatopec absorbed doping products at the Olympic Games in 1948, the molecules lasted only a few hours in his body and thus had a temporary effect on his performance. Other kinds of interventions last longer. Some of them can even be considered as definitively acquired (for instance an implant that is introduced under the skin or a gene that is introduced into the genome of a stem cell). Again, according to this criterion, one can enumerate enhancement interventions in a well-defined order (for the sake of simplicity, I will consider here only two options depending on whether the modification is temporary or definitive, but it is clear that this parameter could also be considered as varying continuously rather than discretely).

Now, the two criteria that we have just identified can be combined. Going through that combinatorial operation, we can define four classes of enhancement techniques that can be presented as in Table 5.1. For each class, examples can easily be found as shown in the table (these examples are not to be confused with the simple paradigmatic example that will be discussed later).

Table 5.1 Four classes of enhancement techniques

	Temporary	Definitive
Localised	Class 1 – Removable prosthesis e.g., running blades and other prosthetic limbs	Class 3 – Implanted prosthesis or localised genetic modification e.g., an implant that is definitively introduced into the body or a genetically modified cell type
De-localised	Class 2 – Doping substance e.g., a substance such as Erythropoietin (Epo) that is absorbed by a sportsman	Class 4 – Germ line genetic modification e.g., a possible genetic transformation that affects all cells of the body (many examples of such modifications could be found in animals, none have been performed in humans, although such modification is easy to imagine)

Ordering enhancement techniques

We have now defined four basic categories (or classes) of enhancement. Next, we will show that these categories can be arranged along a scale that goes from the most artificial to the most natural. In other words, depending on the type of modification being considered, the object (of whatever sort: an apparatus, a gene or a circulating molecule) used to produce the enhancement seems to acquire a status peculiar to the category to which it belongs. In addition, this status seems to vary with respect to the distinction between the natural and the artificial, that is, the transformation can be gauged on a scale measuring its naturalness or artificiality. However, as the object becomes less localised and more definitive, the specificity of its status as artificial or natural becomes less evident.

In the first class, one can find removable prostheses. The distinction between the natural and the artificial here is easy to make: one can identify what belongs to the body and what does not belong to it. The object, although possibly tightly connected to the body, can be put in place and removed, thus making it possible to determine clearly what are the natural and the artificial parts of the ensemble that forms a complete athlete.

In the second class, one can find doping substances such as those that are absorbed by sportsmen in traditional doping techniques. The distinction is a bit more difficult to assess, although not impossible (chemical detection tests might be necessary, for instance).

In the third class, one may find definitively implanted prostheses or, possibly, artificial organs or organs in which a significantly important amount of cells have been genetically modified (for instance, blood stem cells in the bone marrow that have been genetically modified to produce greater amounts of erythropoietin receptors, an example to which I will soon return). Here, we notice that the distinction between what belongs to the body and what does not belong to it is much more difficult to make. Hence, the contemporary discussions and numerous meetings organised by institutions, such as the World Anti-Doping Agency (WADA), which are concerned with issues derived precisely from the difficulty in establishing this distinction, in order to separate prohibited practices from authorised ones.

Finally, in the fourth class, illustrated by a genetic transformation of all the cells of the body (only a theoretical possibility so far), the distinction between the natural and the artificial is nearly impossible to make. Indeed, in this class, the two traditional categories of the natural and the artificial melt into a new category that has not yet received a name. But perhaps we might consider proposing a specific name for such modifications, given their growing importance (possibly 'arti-natural' modifications: that is, artificial modifications that mimic natural ones in such a way that they become indistinguishable from them).

In sum, not only can we establish a distinction amongst classes of modifications, we can also introduce an order relation between the classes that have been defined.[7]

A paradigmatic example for the scale of 'enhancement' techniques

To illustrate the progressive blurring of the distinction between the natural and the artificial as one moves from Class 1 to Class 4 of enhancement techniques as defined above, let me introduce an example that could be taken as a paradigm for the proposed scale of enhancement techniques. The example, in order to be considered as paradigmatic, must cover the whole range of enhancement interventions through a single physiological mechanism. A product naturally present in the human organism that can also be obtained by artificial means, such as erythropoietin, appears to be a good candidate for that.

History and physiology of erythropoietin

Erythropoietin is a hormone that mediates the growth of erythrocytes in the bone marrow.[8] In 1906, Paul Carnot, a professor of medicine in

Paris, proposed the idea that a hormone regulates the production of red blood cells in rabbits. He called this hormone 'hematopoietin'. The postulated hormone was indeed found and later renamed 'erythropoietin' (hormone inducing the proliferation of erythrocytes, also known as red blood cells). Its synthesis by the kidney was demonstrated in 1957. The hormone was isolated and purified in 1977 by Goldwasser and Kung who managed to extract 10 mg of pure erythropoietin from human urine where it is present at a concentration of less than 5 ng per ml (Miyake et al., 1977; Goldwasser, 1994). This extract made it possible to determine the amino-acid sequence of the protein which, in turn, made it possible to identify and then clone the gene coding for erythropoietin (Noguchi et al., 1991). In 1980, the cloned gene was used by the pharmaceutical company Amgen to produce a synthetic form of the hormone that has since been sold under the name 'Epogen'. Epogen was found to be efficient in the treatment of anaemia, which is basically an illness that results from a low concentration of red cells in the blood.

Erythropoietin (Epo) acts by binding to a receptor that is located on the surface of red cell progenitors. The binding of Epo to the Epo-receptor triggers the proliferation and the differentiation of red blood cells. The receptor is itself coded by a gene: the erythropoietin receptor (Epo-R) gene which is located on chromosome 19 in humans. It has been shown that various blood diseases known as erythrocytosis are caused by mutations in this last gene: in these rare conditions, the receptor does not work properly. Detailed studies of these mutations have since led to an important discovery concerning natural genetic advantage carried by some individuals. The history of erythropoietin's physiology, in turn, appears to have been one of the best documented examples of a genetic enhancement that can also be obtained through non-genetic means.

The mutated Epo-R gene as genetic enhancement

In 1991, a group of biologists from Helsinki, Finland, tried to characterise the genetic mutation that causes erythrocytosis. Eeva Juvonen, the first author of this study, was interested in a Finnish family with a high proportion of cross-country skiing winners. The most prominent of these champions was Eero Mäntyranta, who won a gold medal in cross-country skiing at the 1960 Winter Olympic Games in Squaw Valley, USA. He later won two gold medals and one silver medal in the 1964 Winter Olympic Games in Innsbruck and, again, a handful of medals in 1968 in Grenoble. Mäntyranta so heavily dominated the competition that he was suspected of doping. At the time, Epo was already known but it was not available as a doping substance. In the 1990s, Mäntyranta would probably have been suspected of using synthetic Epo for blood doping:

most of the characteristics of his blood were precisely those that permit detection of Epo doping today (including the high density of red blood cells as assessed by the analysis called *hematocrit*). However, although Mäntyranta did occasionally use prohibited substances (amphetamines), he seems never to have used any kind of blood doping. In any case, he did not have to. His advantage came from elsewhere. It came from his own constitution.

Juvonen later managed to determine the kind of mutation that affects Mäntyranta's family. He showed that the champion harbours a single nucleotide mutation in exon 8 (in position 6002 to be precise) of the Epo-R gene that results in a stop codon inside the gene. The translated Epo-R protein is shorter than normal. Interestingly, instead of being less active, the truncated protein product is more efficient than the full length one: the same concentration of Epo in the blood flow results in higher levels of red cells in individuals harbouring the mutation than in their 'normal' counterparts (De la Chapelle, et al., 1991; De la Chapelle et al., 1993).

Juvonen also tried to trace the origin of the mutation. He showed that it presumably occurred around 1855 (about one century before the 1960 Olympic Games), since the first individual identified as a carrier of the mutation appears to have been Eero Mäntyranta's grandfather who was born that year. Ancestors of Mäntyranta's grandfather did not harbour the mutation. The mutation is a dominant one (De la Chapelle et al., 1993), which means that a single copy of the mutated gene in an individual is sufficient to confer the phenotype characterised by the higher than normal propensity to respond to Epo. It does not appear to have any deleterious consequences as judged from its effects on the Mäntyranta family: persons carrying the mutation (about half of the nearly 100 persons who have been studied so far) did not die any younger than the others, nor did they appear more sensitive to illness.

The erythropoietin system as a paradigm for enhancement

Let me recapitulate what is known about the system of red blood cell regulation and its role in performance in sport. Performance in sport is related to the availability of muscular power which, in turn, depends (amongst other things) on the volume of oxygen that reaches the muscles. This volume is directly influenced by the density of red cells present in the blood flow. Red cells are produced by progenitor stem cells in response to Epo. Many actions have been found to be efficient in influencing this system so as to enhance the density of red blood cells in the blood flow:

- The first one, which is not considered an enhancing technique, consists of programming a journey to the mountains. Staying in high altitude for a while enhances the density of red blood cells, because the rarity of oxygen induces Epo synthesis by the kidney which, in turn, triggers the production of red blood cells.
- The second one, the status of which remains problematic for anti-doping authorities, consists of using an oxygen tent (a tent under which the density of oxygen is lower than normal) in order to mimic the effect of a prolonged stay in the mountains.
- The third one, which is clearly considered as doping, consists of injecting synthetic Epo into the blood flow. Again, this will trigger the proliferation and differentiation of red blood cells.
- The fourth method, which would be considered as gene doping if detected (although no case has been reported to date), would consist of the extraction of blood stem cells from the bone marrow of an individual followed by an Epo-R gene transfer into the nucleus of those cells and their reintroduction into the bone marrow of the athlete. Blood stem cells have the peculiar capacity of spontaneously recolonising the bone marrow, thus providing a natural and valuable support for genetic intervention on blood stem cells.
- The fifth method is the one that came about naturally in 1855, presumably in one of the two germ cells that contributed to the embryo of Eero Mäntyranta's grandfather: it consists of mutating the 6002 nucleotide of the Epo-R gene in one of these cells. Such a mutation could be mimicked today by a directed mutation of the same nucleotide in one of the resident Epo-R genes of an embryonic stem cell, later allowed to develop as a full embryo, by using for instance the CRISPR/Cas system.

Thus, we have five methods that lead to a similar physiological result: more red blood cells and consequently more oxygen for the muscles. Now, let us try to locate these various interventions in the four categories of enhancement techniques identified above. The first method can be excluded because it is not an enhancing technique but rather a training technique. What about the remaining four methods? One can easily see that each of them fits into one of the identified categories:

- The use of an oxygen tent is a temporary and localised modification (Class 1 enhancement according to our analysis).
- The use of synthetic Epo is a transient and de-localised modification (Class 2 enhancement).

Table 5.2 The erythropoietin system: a scale of enhancement techniques

	Temporary	Definitive
Localised	Class 1 – Oxygen tent	Class 3 – Genetic modification of stem cells
De-localised	Class 2 – Synthetic Epo injection	Class 4 – Genetic replacement of Epo–R gene

- The genetic modification of red blood progenitor cells corresponds to a localised but definitive modification (Class 3 enhancement).
- The directed genetic replacement of the Epo-R resident gene would be a definitive and de-localised modification (Class 4 enhancement), one that occurred at least once by natural means in Eero Mäntyranta's family.

Table 5.2 recapitulates the analysis. As one can see, it is analogous to the previous one. However, this time, it has been built entirely around the example given by the case of erythropoietin rather than on a theoretical analysis. The concordance of the two analyses allows us in turn to elicit the erythropoietin system as a paradigmatic example of the four-class scale of enhancing interventions proposed above.[9]

Lessons from the example of erythropoietin

The rule already identified in the theoretical analysis also appears clearly in this example: the deeper the incorporation of the modification, the harder it is to maintain the distinction between natural and artificial. Indeed, it is not problematic to assess what is artificial in an oxygen tent. The nature of the intervention can still be assessed when synthetic Epo injections are under consideration. However, the nature of the intervention becomes very difficult to assess when one turns to genetic modifications, particularly when considering point mutations that induce dramatic physiological changes in the whole organism, for these mutations can also occur naturally. Thus, even though the degree of technicality required to insert a point mutation in a cell genome is quite elevated, the result will be indistinguishable from a natural action.

Indeed, the height of the confusion between natural and artificial intervention is reached in the case illustrated by Mäntyranta's family. Here nature appears to have operated like a very skilful sports trainer. If someone were about to undertake the same genetic modification (a point mutation at position 6002 of the Epo-R gene) in the cells of a sportsman who had complained about not having the same natural advantage as Eero Mäntyranta, one might wonder about the kinds of arguments that would be developed to preclude the intervention. Insofar as the intervention appears to be safe (as judged from the natural example given by the Mäntyranta case itself), it would be difficult to claim, as WADA does, that a prohibition of that practice is enforced in order to protect the health of the sportsman.[10]

Thus, ironically enough, the sportsman might claim (as some of them actually do) that it would be unfair not to allow him to enhance himself as much as current scientific knowledge allows, because he would then begin the race with an identified handicap.[11] The irony here comes from the fact that fairness is a notion that is commonly used to ban doping techniques whereas it is here turned into an argument in favour of such techniques. Intervention involving the introduction of a point mutation in a given gene would thus challenge the conceptual framework that has been elaborated to justify the current policy on doping.[12] We should not be surprised, however, that once the distinction between the natural and the artificial becomes totally blurred, the questions that this distinction has helped to solve return to the fore.[13] This time though, there would no longer be a couple of opposing concepts (such as natural and artificial) that would help ground a prohibiting principle in order to justify a regulatory system. This situation appears to take us a step further along what Peter Conrad called a 'medicalization of society' (Conrad, 2007; Simonstein, 2008).

Anticipating the effect of a modification

Let us imagine a sport assistant (a profession that is likely to develop widely in the coming years)[14] who is searching for techniques to enhance an athlete's respiratory capacities. Let us suppose that she knows the system of Epo regulation well. She might think about the oxygen tent or about Epo injections to enhance red blood cell density. If she has some training in genetics, she might even think about introducing an Epo gene into kidney cells in order to enhance the athlete's endogenous production of Epo. If she does so, she would be acting like an engineer or a designer in charge of a complex machine who has gained access to several basic commands and activates some of them, but who nevertheless still has a poor idea of the complete consequences of what she is doing.[15] Even a well-informed and trained biologist would not be able to predict the full consequences of most of the genetic interventions that can be considered. She would only recall that, in many instances, genetic engineering remains a hazardous procedure. As Lee Sweeney, a scientist who created a transgenic mouse with enhanced muscular capacities, put it in a *Scientific American* article: 'it is clear that gene doping can be dangerous and detrimental to health. The case of gene doping is special in the sense that there are many uncertainties as to the long term effect' (Sweeney, 2004).[16] Thus, entering into such a doping strategy is somewhat risky. There is, however, a much less risky and even presumably safe procedure (although perhaps quite a laborious one): introducing a point mutation in position 6002 of the Epo-R gene in red cell progenitors located in the bone marrow of the athlete. This would mimic what nature had done in the case of Mäntyranta. Such an intervention would hardly be imagined by an engineer for its logic could not be discovered unless one introduces a historical (or evolutionary) perspective that is not often found in engineering reasoning (Gerdes, 2006; Pound, 2009). In other words, the logic of the intervention would here be given by the observation of nature. Should it be called natural or artificial? Perplexity appears to be legitimate.

Conclusion

The current enhancement debate, mostly organized and moderated by the so-called 'transhumanist' philosophers, often turns to an examination of whether future interventions in humans are likely to resemble those of designers of living matter (see, for instance, *Designer Evolution* (2006) by Simon Young, an influential book on that topic).[17] The human body has its own objections to such projects: even when newly born, it is, indeed, a very old object. It encapsulates a very long history that looks young only because it is renewed again with each generation. Consequently, the project of enhancement may remain hazardous as long as it is conceived from the point of view of an engineer.[18]

Does that mean, in turn, that attempting to enhance the human body to achieve better performance is always dangerous and should be banned according to some precautionary principle? Does that mean that the attempts to modify the human body only represent a hubris-like action that would lead, sooner or later, to some kind of Promethean punishment?[19] The century-and-a-half-old lesson from Darwin actually suggests another idea of what enhancing the performance of the human body can mean.[20] For one of the ideas that has emerged from the work of Charles Darwin is that the evolutionary process does not resemble the work of an engineer. According to modern biological thought, the human organism has been built through millions years of evolution. As a consequence, one can certainly enhance certain aspects of its performance. But there are chances that this will lead to discrepancies that may appear only in the long run. The example of the erythropoietin system could here provide a basis for a methodical approach to human enhancement: instead of going from the available technique to the living organism, this approach would proceed from the study of contemporary living organisms and try to determine where the better performances of some of them come from, taking as a rule that the way a given advantage is provided to an organism is highly unpredictable. These interventions are both natural (in that they do imitate nature) and artificial (since they are produced by humans). We could call them 'arti-natural'.

Thus, the origin of the particular symptom that I mentioned in the introduction and that can be characterised by the tendency to look at humans as limited organisms (rather than as organisms with a potential), might reside in the propensity to look at living organisms as designers instead of looking at them as evolutionists. And a cure for this 'illness' may be, at least in part, a conceptual one: defining a concept of interventions that could be labelled 'arti-natural', meaning interventions that would be neither natural nor artificial and that would mimic through artificial means natural modifications previously found and characterised in humans.

Notes

1. The current body of work on human enhancement is growing abundantly. The most cited works among recent publications include Miller and Wilsdon (2006), Moore (2008) and Savulescu and Bostrom (2009).
2. Druckman and Bjork (1991) provided an early analysis of this topic.
3. For a general introduction to this debate, see, for instance, Maxwell and Bailey (2005) or Naam (2005).
4. Elaborate attempts to define the concept of enhancement can be found, for instance in Savulescu and Bostrom (2009).
5. Early and classical papers on the question of human enhancement can be found in Swets (1988).
6. A discussion of doping techniques as well as a catalogue of such techniques presented in a historical perspective can be found in Rosen (2008).
7. Further analysis on the same topic can be found in: Schneider and Friedmann, 2006.
8. For background on erythropoietin, see Jelkmann (2007), Jelkmann and Gross (1989), Goldwasser (1996) and Goldwasser (1991).
9. The measure of the efficiency of Epo for the treatment of anaemia has been largely documented through standard double-blind techniques. The same techniques could obviously be used to assess the efficiency of Epo as an enhancement procedure in sports. However, due to the prohibition of the technique, the efficiency of this doping method relies only on occasional data. Although these data are often very clear ('*you can surely climb the Mont Ventoux with or without Epo, but I can tell you one thing: if you race once with Epo, you will never do it again without Epo*' stated a rider), it remains difficult to assess its efficiency on a quantitative basis.

10. See the official document published by WADA, *World Anti-Doping Code*, 2009, inside front cover: 'The World Anti-Doping Code was first adopted in 2003 and became effective in 2004. It incorporates revisions to the World Anti-Doping Code that were approved by the World Anti-Doping Agency Foundation Board on November 17, 2007. The revised World Anti-Doping Code is effective as of January 1, 2009.'
11. An eloquent defense of human enhancement is provided by Agar (2005).
12. A discussion on rhetorical justifications of anti-doping policies is given by Mitchell (2007).
13. For an ethical and social perspective on these questions see Parens (1998).
14. A similar scenario was examined by others after seminal gene doping papers were published. See, for instance, Brownlee (2004), Murray (2003) and Skipper (2004).
15. Further discussion on the dialectics of knowledge and ignorance in attempts to enhance human performance can be found in Zonneveld et al. (2008).
16. In 2002, Lee Sweeney published a paper in which he describes a dramatic enhancement of muscular capacities in mice obtained after a genetic modification. See Barton et al. (2002).
17. See also Zonneveld et al. (2008).
18. For a thorough discussion of the hazards associated with gene doping, see Schneider and Friedmann (2006).
19. For an analysis of the discourses of opponents to human enhancement, see Gillespie et al. (2006). See also Jason (2005).
20. This view has often been defended. See, for instance, Tomasini (2007) and Weckert (2007).

References

Agar N. (2005) *Liberal Eugenics: In Defence of Human Enhancement* (Malden, MA: Blackwell Pub).

Aristotle (4th century BC). *Physics*, translated by Robin Waterfield, with an Introduction and Notes by David Bostock (Oxford: Oxford University Press, 1996).

Barton E., Morris L., Musaro M., Rosenthal N. and Sweeney L. (2002) 'Muscle-specific expression of insulin-like growth factor 1 counters muscle decline in mdx mice', *Journal of Cell Biology*, 157 (2002), 137–48.

Bensaude-Vincent B. and Newman W. (2007) *The Artificial and the Natural: An Evolving Polarity* (Cambridge, MS: MIT Press).

Brownlee C. (2004), 'Gene doping: Will athletes go for the ultimate high?', *Science News* 166: 280–81.

Canguilhem G. (1989) *The Normal and the Pathological* (New York: Zone Books). [First French edition: Paris, Presses Universitaires de France, 1966.]

Conrad P. (2007) *The Medicalization of Society: On the Transformation of Human Conditions into Treatable Disorders* (Baltimore: Johns Hopkins University Press).

De la Chapelle A., Sistonen P., Lehväslaiho H., Ikkala E. and Juvonen E. (1993) 'Familial erythrocytosis genetically linked to erythropoietin receptor gene', *Lancet*, 343(8837), 82–4.

De la Chapelle A., Träskelin A.L. and Juvonen E. (1991) 'Truncated erythropoietin receptor causes dominantly inherited benign human erythrocytosis', *Proceedings of the National Academy of Sciences USA*, 90(10), 4495–9.
Druckman D. and Bjork R.A. (1991) *In the Mind's Eye: Enhancing Human Performance* (Washington, D.C: National Academy Press).
Gerdes L.I. (2006) *Humanity's Future* (Detroit: Greenhaven Press).
Gether C. and Hirst D. (2009) *Damien Hirst* (Ishøj, Denmark: Arken Museum of Modern Art).
Gillespie N., Bailey R., Cohen E. and Garreau J. (2006) 'Who's Afraid of Human Enhancement? A Reason debate on the promise, perils, and ethics of human biotechnology', *Reason*, 37(8), 22–33.
Goldwasser E. (1991), 'The biology of erythropoietin', *Blood Purification*, 9(3),119–22.
Goldwasser E. (1994) 'The oxygen sensor and erythropoietin gene regulation', *Annals of the New York Academy of Sciences*, 718, 1–2.
Goldwasser E. (1996) 'Erythropoietin: a somewhat personal history', *Perspectives in Biology and Medicine*, 40(1), 18–32.
Jason R. (2005) 'Human Dispossession and Human Enhancement' *American Journal of Bioethics*, 5.3, 27–9.
Jelkmann W. and Gross A.J. (1989) *Erythropoietin* (Berlin: Springer-Verlag).
Jelkmann W. (2007), 'Erythropoietin after a century of research: younger than ever', *European Journal of Haematology*, 78(3), 183–205
Juvonen E., Ikkala E., Fyhrquist F. and Ruutu T. (1991), 'Autosomal dominant erythrocytosis caused by increased sensitivity to erythropoietin', *Blood*, 78(11), 3066–9.
Maxwell M.J. and Bailey R. (2005) *Perfecting People through Biotechnology: The Implications of Human Enhancement for Society* (Atlanta, GA: Georgia State University Center for Law, Health & Society).
Miller P. and Wilsdon J. (2006) *Better Humans? The Politics of Human Enhancement and Life Extension* (London: Demos).
Mitchell C.B. (2007) *Biotechnology and the Human Good* (Washington, D.C: Georgetown University Press).
Moore P. (2008) *Enhancing Me: The Hope and the Hype of Human Enhancement* (Chichester, England: Wiley).
Miyake T., Kung C.K. and Goldwasser E. (1997) 'Purification of human erythropoietin', *Journal of Biological Chemistry*, 252(15), 5558–64.
Murray T. (2003) 'An Olympic tail?', *Nature Review Genetics*, 4: 494.
Naam R. (2005) *More Than Human: Embracing the Promise of Biological Enhancement*, (New York: Broadway Books).
Noguchi C.T., Bae K.S., Chin K., Wada Y., Schechter A.N. and Hankins W.D. (1991) 'Cloning of the human erythropoietin receptor gene', *Blood*, 78(10), 2548–56.
Parens E. (1998) *Enhancing Human Traits: Ethical and Social Implications* (Washington, D.C: Georgetown University Press).
Pound R.W. (2009) 'Human Enhancement', *Issues in Science and Technology*, 25, 4: 5.
Rosen D.M. (2008) *Dope: A History of Performance Enhancement in Sports from the Nineteenth Century to Today* (Westport, Conn: Praeger).
Savulescu J. and Bostrom N. (2009) *Human Enhancement* (Oxford: Oxford University Press).

Schneider A.J. and Friedmann T. (2006) *Gene Doping in Sports: The Science and Ethics of Genetically Modified Athletes* (Amsterdam: Elsevier Academic Press).
Simonstein F. (2008) 'Human Enhancement and Factor X', *Journal of Medical Ethics*, 34, 2: 102–3.
Skipper M. (2004) 'Gene doping: A new threat for the Olympics?' *Nature Reviews Genetics*, 5: 720.
Sweeney L. (2004) 'Gene doping', *Scientific American* (July), 63–69.
Swets J.A. (1988) *Enhancing Human Performance: Issues, Theories, and Techniques; Background Papers* (Washington, D.C: National Academy Press).
Tomasini F. (2007) 'Imagining Human Enhancement: Whose Future, Which Rationality?' *Theoretical Medicine and Bioethics*, 28, 6: 497–507.
Weckert J. (2007). 'What's Wrong with Human Enhancement?' *Australasian Science*, 28(9), 16–17.
World Anti-Doping Agency (WADA). (2009) *World Anti-Doping Code*. Available at: http://www.wada-ama.org/Documents/World_Anti-Doping_Program/WADP-The-Code/WADA_Anti-Doping_CODE_2009_EN.pdf . Accessed on November 9, 2013.
Young S. (2006) *Designer Evolution: A Transhumanist Manifesto* (Amherst, N.Y: Prometheus Books).
Zonneveld L., Dijstelbloem H., Ringoir D. (2008) *Reshaping the Human Condition: Exploring Human Enhancement* (The Hague: Rathenau Institute).

6
From Repair to Enhancement: The Use of Technical Aids in the Field of Disability

Myriam Winance, Anne Marcellini and Éric de Léséleuc

Introduction

There is nothing new in the use of technology to repair and compensate for human disabilities. Throughout the centuries we find examples of prostheses and artificial limbs being used to replace lost limbs (Avan et al., 1988). Various technical aids were also used to compensate for the body's failings or to facilitate treatment – wheeled vehicles were used to carry invalids, for example. During the 16th and 17th centuries, the first wheelchairs that could be propelled by the users themselves appeared. But the majority of these vehicles were made of wood; they were heavy, cumbersome and difficult to manoeuvre. During the 19th century, medical progress (the discovery of anaesthesia, asepsis, antibiotics, radiology, and so on) made it possible to develop new techniques to repair and compensate for deficiencies. Furthermore, the end of the 19th century saw the beginning of a change in the social treatment of disabled persons, leading to the emergence of the notion of 'handicap' as a replacement for the notions of infirmity, invalidity, idiocy, and so on. In other words, developments in the modes of repairing deficiencies and of compensating for disabilities correlate with changes in the definition of 'disability' (as 'handicap') and in the way persons with disabilities are integrated into society.

In this chapter, we wish to focus on the history of this correlation and analyse the ways in which repair and compensation have been handled throughout the 20th century and are being handled today. We will examine and compare compensation in everyday life and in sport, by addressing two issues: that of the social legitimacy of repair and compensation, and that of people's experiences – their relationships with technical aids. In both areas, we will analyse the challenges and the objectives of the processes of repair and compensation for deficiencies and disabilities.[1] In both areas, the objective is to acquire the capacity 'to do', but what this means differs according to the area: in the context of everyday life, the normative objective is 'to be able to do what the average person can do'; in the context of sport, the normative objective is to go beyond average capacities. As we will see, this process of acquiring capacities is based on the process of the body adjusting to the technical aid and on the process of coming to a practical arrangement with one's human or physical environment – processes which are then hidden, made invisible by social attitude. By thus examining the ways in which people with impairments use technical aids – uses that create a tension between an impairment-reduction logic and a capacity-enhancement logic – this chapter aims to shed light on the subject of the enhancement of the human body and related ethical issues.

The everyday world: acquiring the capacity 'to cope'

From rehabilitation to accessibility: being and doing the same as everyone else

Rehabilitation practices began to take shape at the beginning of the 20th century. They contributed to a great change in the representations and practices regarding people with impairments, an evolution which led to the notion that deficiencies and impairments must be repaired, both physically and socially. This led, in the 1950s, to the use of the terms 'handicap' and 'handicapped persons' to designate people with impairments. The notion of 'handicap' designated a deviation from the social norm (defined in terms of social performance), that was the consequence of a deviation from the bodily and functional norm.

H.J. Stiker (1997) dates the beginning of this transformation at the end of the 19th century, with the issue of accidents at work. In France, this issue was resolved by law in 1898 (Ewald, 1986). It set out the principle of *social repair*, in the shape of financial compensation for the damage caused by an accident at work. The First World War continued this approach by extending the same right to disabled ex-servicemen, for whom a system of disability pensions was created. However, the First World War also caused a shift. Due to the lack of workers, it became necessary to reintegrate the disabled ex-servicemen into the workplace. This meant developing a rehabilitation system that was then extended to all disabled civilians. From a social standpoint, the purpose of these practices was to allow people to return to work; from a medical point of view, the objective was to reduce impairments and to restore all functions to the body, so that people could once again live an ordinary life. This was achieved with prostheses and ortheses and by muscular exercises. It was a case of doing what able-bodied people did, *in the way that they did it*. Rehabilitation and functional re-education were designed to bring disabled persons in line with the model of the able-bodied, both in terms of social aptitude and with regard to physical and functional skills. For survivors of poliomyelitis (Wilson, 2009) and for paraplegics, the purpose of rehabilitation was to learn to walk again and not to acquire a capacity to move around; recovery of the capacity to walk meant one was cured and had returned to normality, whereas the need for a wheelchair signified failure and 'definitive impairment'. The objective was thus alignment, in the strongest sense of the word, because rehabilitation was aimed not just at the acquisition of average ordinary capacities, but also at the similarity of appearance. This pursuit of 'visible normality' guided the changes made in prostheses, designed so as to resemble the limb that they were replacing, in colour, shape and even texture.

The 1930s saw the creation in the United States of the ancestor of our current wheelchairs, the manual foldable wheelchair, a lightweight chair with a tubular iron structure. The purpose of the wheelchair was no longer to repair the body but to compensate for the loss of mobility. In describing the experience of paraplegic Canadian veterans, M. Tremblay (Tremblay, 1996; Tremblay et al., 2005) describes how the shift from pendular walking to wheelchair use had represented, for them, the opportunity to leave hospital and return to mainstream society. To achieve this, they adapted to the existing environment (by finding accommodation with no stairs, and so on) and relied on help from other people (to carry them when they were unable to avoid stairs). A wheelchair enabled the person to live as part of the community. It granted them social and functional normalisation, without directly normalising their bodies and without making them 'standing persons'. In France, the abandonment of pendular walking as a means of movement[2] and the more systematic recourse to wheelchairs is probably more recent, dating back to the 1960s or 1970s.[3] In both cases, the use of manual wheelchairs as a means of movement marked a shift in attitude: whereas the objective was still to acquire normal capacities, this no longer involved an alignment with ordinary physical and functional capacities but the acquisition of new functional capacities pertaining specifically to the use of the wheelchair.

In the 1970s and 1980s there was another shift under the impetus of the disabled people's movements created at that time (Barton and Oliver, 1997; Campbell, 1997; Oliver and Barnes, 1998). These people – and wheelchair users in particular – were regularly confronted with obstacles in their environment: unsuitable workplaces, a lack of accessible and adapted housing, and so on. They became aware that they were not disabled because of their impairments, but because they were excluded from an inaccessible society. So instead of adapting themselves to that society, they asked society to change.[4] On the basis of their experience of exclusion, disabled activists developed what is known as the social model of disability. It states that disability is not the result of an impairment, but of the obstacles (physical, cultural, and so on) set up by society that prevent disabled persons from participating. Within the social model, the objective is still to allow people to acquire ordinary capacities[5] making it possible for them to live in 'ordinary' society. But by reversing the causal process of disability, the social model also changes the process through which these capacities are acquired. It takes for granted the existence of a variation in humankind and places the weight of normalisation on society's shoulders. In other words, the starting point is no longer 'the person who cannot walk', but 'the person in a wheelchair', to whom society must adapt.

During the 20th century, the field of disability was constructed through reference to a logic of normalisation in terms of an alignment with the 'able-bodied' person; the extent of this alignment has varied over the years (integrating appearance and/or functional capacities) and taken different concrete forms: the rehabilitation of the individual or adaptation to the environment. Recourse to technical objects – ortheses, prostheses and technical aids – was central to this normalisation process. In the field of rehabilitation, technical aids were used to repair the body and to compensate for incapacities. The purpose of the work done by professionals was to improve the disabled person's interaction with technical objects. Conversely, the social model takes, as the starting point for its logic,[6] people as they are – in wheelchairs, with walking sticks, with hearing aids – without worrying about their relationships to technical aids. In the next section, we will describe the work through which people's abilities are defined, by looking at the interaction between people and their wheelchairs.

Living and making do with a technical aid: adjusting, accommodating, making arrangements

The wheelchair is one of the most ordinary technical aids. It is used by a wide variety of persons in terms of sex, age, type of disability, place of residence (ordinary or institution), and so on (Vignier et al., 2008). To this diversity of users corresponds a diversity of uses: use may be temporary or permanent, partial (just for certain activities) or total (all day long). But as Mr Doris, a paraplegic who has had one limb amputated, says:[7] 'It's true that it is easier to move around on two legs than on four wheels, but when you don't have a choice, you have to do it' (December 13, 2007). What does *do* mean here? Having to use a wheelchair means no longer moving around on two legs and having to experience a different body (seated, maybe partially paralysed and numb, or maybe stricken by uncoordinated movements). It means moving around on four wheels and living in, with, through a technical object that people normally do without. Hence a process of adaptation that has two facets: a process of adjustment and a process of accommodation.

The adjustment process refers to the mutual and planned adaptation between person and chair. This process can be seen when people are choosing and trying out a new wheelchair. The question that guides the wheelchair trial is: 'What are you going to do with your chair each day?' (Interview with an ergotherapist in charge of tests, December 2007). To provide an answer to this question, the wheelchair trial must take on the form of a joint analysis of the disabled person's situation. The idea is to explore how they feel and what they will be able to do 'in this wheelchair' on an everyday basis. This exploration goes hand in hand with work on transforming one's perceptions and possibilities for action. On the one hand, people are confronted with their perceptions in each chair they test and, on the other hand, their perceptions are modified by gradual changes to their position (position of their arms, legs, and so on) and to the wheelchair itself[8] (adjustments to the back, the arm rests, the footrests, the addition of a head rest or of a cushion, and so on). This adjustment process is reflexive, in as much as it involves putting a distance between the subject, their body and their wheelchair. The purpose of this process is to discover the position in which people feel 'comfortable' and to find them new abilities or possibilities for action. Through this process, which continues in day-to-day life, people's experiences – their perceptions, opportunities for action, mobility range, social space, and so on – are gradually shaped.

The process of accommodation (Thévenot, 1994) takes place both during and after the adjustment phase; it involves the non-reflexive adaptation to and of the wheelchair, the 'material' shaping of the body and chair. The more people use their wheelchairs, the more they become used to it, in terms of their bodies, their positions and their ways of acting. Mrs Debra's story illustrates this process. Mrs Debra, aged 57, has problems walking long distances. She uses a wheelchair for group outings. She explained to me [Myriam Winance] that she felt that her footrests were too high, so she had removed them; when she used them, her legs were too bent. At the end of the interview, I asked if I could see her wheelchair and she agreed. It was a standard model. She once again explained that she did not use the footrests and she sat in the chair to show me that they were not suitable. Her legs were indeed a bit high. I looked at the footrests and saw that they could be lowered a notch. I explained this to her; she hesitated and I offered to adjust them for her. She hesitated again and then told me:

> 'But aside from that, what I'm saying is... you see, I'm sitting like this, you see for example, ... I use my feet ... so ... and then ... it takes up a lot of room, be careful ... because in a minibus for example, sometimes, when there are one or two other persons, well, ... it's a bit cramped ... no, what I'm saying is that I don't really need footrests, they get in the way when I want to move my legs forward [that is when she stretches them out in front of her] ... I don't know ... maybe I will adjust them, maybe I won't adjust them because ... [she removes them and places them in a corner of the garage]' (September 5, 2008).

In this example, we first see the adjustment process; Mrs Debra finds the footrests to be a nuisance, so she removes them from the wheelchair. She thus feels more comfortable when seated. But when she is being pushed, she is obliged to hold her legs straight out in front of her, which does not seem to be very comfortable. Yet she finally becomes accustomed to this position and, when she is offered an alternative solution, she is very hesitant about changing. Adjustment continues through the process of accommodation, that is the incorporation of a position initially experienced as uncomfortable, but to which the person adjusts because the position finally appears more practical. And this process implies not only an adjustment of the person to the chair, but also of the person in the chair to their usual environment. In Mrs Debra's example, the removal of the footrests is also an adjustment with regard to the minibus and to the lack of room when there are several people. A person's abilities are not only the result of the dual processes of adjustment and accommodation, but also of the practical arrangements that people make in order to be able to go about their daily activities.

These practical arrangements involve a process of adaptation between wheelchair users and their environments, in accordance with the human and non-human resources available within the latter. The different ways in which people go shopping provide a good illustration of this process.

For example, Mr Doris, who is very much at ease when using his wheelchair, chooses to do his shopping in medium-sized supermarkets, preferring to avoid hypermarkets, despite the fact that both are accessible by wheelchair. However, when he does his shopping, he has to operate his wheelchair with one hand and push the trolley with the other. This is not easy because, when the trolley is full, it rolls all over the place. So he cannot push it over long distances. Because he wishes to cope on his own, he has therefore opted to do his shopping in medium-sized supermarkets, where the distances are better suited to his 'shopping mobility'. Mrs Lepetit, who suffers from amyotrophic lateral sclerosis and who uses a wheelchair on a permanent basis, no longer does her own shopping, letting her husband do it on his own. Mrs Pichard, who is hemiplegic and who currently uses a standard wheelchair with a simple handrim,[9] accompanies her husband into hypermarkets; they use a special trolley which attaches to the wheelchair, with Mr Pichard pushing the whole unit. The latter example shows that practical arrangements depend on the human and non-human resources available in the environment of the person in question (in this case, special trolleys and the husband's assistance). In the same way, when buying his medication, Mr Doris has set up an arrangement that relies on the human resources available to him in his environment:

> *Mr Doris*: Making everything accessible is a great idea, but you mustn't get carried away [The pharmacist had some work done on his shop to make it accessible]. They installed a ramp inside the shop;...due to the lack of space they had to make a spiral ramp...there was quite a height to go up and the ramp took up about 4 square metres in the shop. Was it really worth it? If I need medication, I give them a ring and then I stop outside the shop...it's easier for the pharmacist to bring me the medication than it is for me to get out of the car....When I go there every month, there's a lot to cope with – what with all my stuff and the fuss with all the full bags. If I had to get out of the car, go into the pharmacy and especially come out again with packages, how do I manage if I have my hands full? How can I push my wheelchair when I'm on my own?...They do have a ramp. But in any case it's not always easy to park. It's one thing to park on the road. But then to get the wheelchair out with all the traffic going by, it's not easy (December 13, 2007).

This example demonstrates the difference between accessibility and real mobility – activity.[10] Accessibility requires the physical and social environment to be adapted to individual specificities. But it never covers all individual specificities. Hence the need for practical arrangements that reduce the gap between opportunities for action and actual actions through a process of adjustment. This is what can be seen in the above example. Although the pharmacy is now accessible, it is still difficult to get in and out; indeed for Mr Doris it is potentially dangerous as, with all the traffic, he cannot easily transfer from the car to the chair and, when leaving the pharmacy, both carry his packages and operate his wheelchair. He has therefore come up with an alternative, based on anticipation and relying on the resources available in his social space – in this case his pharmacist, who is prepared to leave her shop. For the 'person-in-a-wheelchair', because the action is the result of practical arrangements, it has to be thought out and organised in advance.

In a given situation, the extent to which these anticipated practical arrangements can be adapted or avoided depends in part on the individual's expertise in using a wheelchair, and in part on the resources on which they can rely in order to organise new arrangements. Let us imagine a wheelchair user who is alone, and who regularly takes the Paris Metro to travel to the Forum des Halles. The Châtelet–Les Halles station has a lift, but it is sometimes out of order. In such a situation, either the person can go no further or they have acquired enough skill to use the escalators on their own or else they ask for (or passers-by offer) help in using the escalators or stairs. In the latter case, as Quéré says (Quéré and Relieu, 2001), the offer or request for help puts the wheelchair user in a situation of visibility; this breaks the vagueness and anonymity that govern relations within public space. Finally, depending on the practical compromises which have been made in each case, what the person is will vary: they will be either 'a-mobile-person (-in-a-wheelchair)', or 'a-person-in-a-wheelchair'. In the first instance, the wheelchair becomes an integral part of that person; in the second instance, it is a world object in relation to which the person must either act or be acted upon, as shown by the following extract:

> *Sociologist*: You never have anyone push you?
> *Mr Doris*: Oh no. I hate that. I hate that. Sometimes when I'm with other people, able-bodied people always think they are helping me by pushing me, but... It increases my impression that I'm disabled. When I push myself, I move around, I go wherever I want. It's true that I'm disabled but I don't feel it. Whereas if someone pushes me, and then... It's always the same... imagine you are somewhere where there are lots of people, the person who is pushing you

doesn't have the same feeling as the person in the wheelchair, you know, someone who is pushing does not look too carefully. Whereas me, I've always... I have to steer, I've always got both my hands on the handrims to avoid hitting someone.... I can judge distances better, I can move fairly quickly in the middle of a sizeable crowd, ... I can stop, I can... it's automatic, a little touch on the rim, I go right, I go left and I avoid the obstacle (December 13, 2007).

Gradually, with regular use, people get to know their wheelchair, feel its reactions and incorporate it. The person no longer merely uses (actions) the wheelchair; the chair becomes that through which they act. The chair becomes easy to use, indicating a oneness of perception and action. The wheelchair truly becomes part of them. At this point, the process of adjustment and of accommodation has become invisible, imperceptible. It is concealed by action which unfolds naturally. But the wheelchair is not always 'my legs'; it can also become 'my disability'. As Mr Doris notes, the simple fact of being pushed makes him feel disabled; in other words, it changes the status of the wheelchair, from that of 'his legs' to 'a wheelchair in which he is pushed'. The capacity to fully incorporate the wheelchair can be called into question at any moment by certain situations: a breakdown, a pain, getting old, putting on weight or even the environment. Moving around becomes more difficult or even impossible, reminding persons that they are in a wheelchair. A disconnection occurs: people no longer act through their wheelchairs but on them, needing to think about how to handle them or to have them handled, in order to get to a given place. Bodies with prostheses are feeble bodies: their abilities are shaped by a process of adjustment and accommodation that can be disrupted at any moment, causing the disability and the sensation of being disabled to emerge.

The capacity to cope with everyday life thus results, on the one hand, from the expertise acquired during the process of adjustment and accommodation to the wheelchair, and, on the other hand, from the implementation of practical arrangements. These practical arrangements rely upon resources (human and non-human) offered by the environment, hence the importance of accessibility, in terms of buildings and services. Finally, the status of the wheelchair, as an incorporated or a world object, varies not only according to the process of adjustment/accommodation, but also to the situation of its use.

Functional rehabilitation practices, the use of technical aids and the accessibilisation of society thus aim to improve the physical, functional and social situation of disabled persons, albeit an improvement that is limited to the objective of 'autonomy in everyday life' and 'the return to normal life': the objective and end point are an average value. 'Repair' and 'compensation' aim to increase the range of abilities of disabled persons, but always within the framework of a project delimited by a normative reference – that of 'ordinary activities'. Indeed, the logic of improving capacities and the conditions of life necessary to such improvement acquires its social legitimacy through its reference to the ideology of the fight against inequality, the aim of which is to align the 'unfit' with the average able-bodied person. Conversely, the world of sport, and, more generally, contemporary society, sustain a logic of perpetually improving performance and capacities.

The sports world: always try to surpass your performance

'Disabled' sportspeople: from rehabilitation to pushing one's boundaries

Rehabilitation through sport is a logical way to develop the capacities of people with disabilities, until they reach 'normal' capacities. But there is already a difference when one goes from rehabilitation through sport to Paralympic sport, which requires the mindset of permanently pushing one's boundaries. Over the second half of the 20th century, the world of sport was divided into two domains corresponding to two categories of people: one reserved for able-bodied people and the other for disabled persons. This separation was made on the basis of two criteria: the biological impairment which affects certain persons and the inferior sports performances which supposedly result from this. This also led to another difference between the two categories of sportspeople – the role of technical aids. In Olympic sports, athletes compete 'naked' and are not allowed to use technical aids. On the other hand, in Paralympic sports, which bring together sportspeople with different types of disability, recourse to technical aids is legitimate and difficult to challenge. Technical aids are not only tools that people use to recover their normal capacities (that compensate for incapacities caused by impairments), but also tools that allow athletes to develop their greatest possible capacities and performances. Research on technical sports aids is thus developing rapidly, with the development of ground-breaking technology that aims to constantly push the boundaries of existing performances. Indeed, in sport, be it Olympic or Paralympic, the objective of normalisation is replaced by the objective of surpassing what are considered to be the human body's 'natural' capacities. It is no longer a question of becoming 'ordinary' but of aiming to become 'extra-ordinary'. Within this framework, which goes beyond any logic of rehabilitation, the performances of disabled athletes will gradually improve, but on the basis of a sporting principle that sets no *a priori* limits. Only one symbolic barrier remains, that of a hierarchical ordering of athletes according to their initial 'natural' aptitudes, implying that people who are physically impaired cannot put in better performances than people who have no impairment.

Sportspeople with prostheses: the logical irruption of endless disorder, or the destabilisation of the hierarchy of 'natural qualities'

Recently, the case of one particular athlete, Oscar Pistorius, arose to challenge the established distinctions between 'able-bodied sport' and 'disabled sport': a bilateral tibial amputee who uses prostheses – 100 per cent carbon fibre blades ('Flex-Foot Cheetah') – to run, was on the point of equalling the performances of able-bodied athletes. The controversy broke out during the Paralympics in 2004. Pistorius, aged 18, won the 200 m gold medal and the 100 m bronze, racing against veteran runners with single tibial amputations. The controversy exploded when Pistorius, whose performances were improving – in the 2007 'able-bodied' championships held in South Africa, he had finished second in the 400 m race –, asked to be allowed to take part in international athletic competitions. The entire debate revolved around the role of his prostheses, which were claimed to represent a potential advantage for his performances. During this debate, Oscar Pistorius became the human who had been improved by technology. The sports world was to qualify this supposed 'artificial' improvement to his results by a 'technical aid' as 'technodoping'. Indeed, in sport, 'real doping' is that which alters a being, which changes his/her essential identity, which modifies the 'natural' biological identity of a human, and which in so doing ruins the sporting objective of setting the hierarchical order of mankind's 'natural' value.

This aspect of the essential 'naturality' of champions is explained by Dr. Giuseppe Lippi, a specialist in clinical biochemistry, haematology and doping at the University of Verona. In March 2008, in the midst of the Pistorius controversy, he co-authored an article entitled 'Pistorius ineligible for the Olympic Games: the right decision' (Lippi and Mattiuzzi, 2008). For the authors, '[a]thletic performances (and champions) are largely genetically determined and genes are the product of natural selection. Technology is a great aid and the most favourable opportunity to overcome disabilities in daily life. It has nothing to do with traditional competitive sports, however, especially if 'cyborgization' is challenging to replace nature's own evolutionary scheme'. The article clearly asserts the legitimacy of technical aids in 'overcoming physical disabilities in everyday life', whilst at the same time denying any such legitimacy in the world of 'traditional sport'.

It was indeed the 'technologically hybrid' nature of Pistorius, equipped to run, that was being called into question in the controversy surrounding the athlete, in as much as this nature seriously disturbs the sports logic of comparison and hierarchical ordering of natural human capacities. Hence Pistorius' battle to break out of the 'doped athlete' category and enter the category of champions. This battle involved an administrative project to prove that his sporting performances were due to his 'qualities', to his natural aptitudes and not to his prostheses. In other words, he had to show that his performances were 'his own' and not linked to the technical device that he had used to run. To achieve this, he stressed his 'personal qualities': 'This type of prosthesis has been used for 14 years by other athletes who have never achieved my results. Which proves that my performances are due to my talent and to my training' (Hirsch and Mathiot, 2007).

The Oscar Pistorius case is a major problem for the sporting institution, because it combines two issues in a way never before seen: the production of a performance potentially superior to that of able-bodied athletes and the fact that the performance was produced by using a prosthesis that replaces a part of the athlete's body. In the history of Olympic sport, there have of course been occasions in which Paralympic athletes were accepted as competitors: in 1984, Neroli Fairhall, a wheelchair athlete, came 35th in archery at the Olympic Games. Marla Runyan, a partially sighted American runner, took part in the Olympic Games in 2000 and 2004. In 2000, Terence Parkin, a deaf swimmer, won the silver medal in the 200 m breaststroke. But all of these 'Olympic athletes with impairments' competed 'naked'; none of them used any type of technical aid to compensate for their disability.[11] The sporting institution could therefore consider them to be legitimate. They did indeed have certain 'anomalies' or impairments of their biological bodies but, biologically speaking, they could nonetheless be considered as 'pure' and 'natural' athletes. Whether the 'anomaly' was a difference in the biological body considered as an 'impairment' when compared to theoretical organic integrity, or a difference considered to be a biological 'peculiarity' providing an advantage compared to the norm (such as the difference in the size of Usain Bolt's calcaneum, which supposedly explains his extraordinary performances), in both cases, what makes sense in sports logic are the diversity in the biological realities of the human body and the differences in performance that such diversity entails. The runner Oscar Pistorius describes himself as a problematic 'mixture' of man and machine. [12]

Unlike other Paralympic athletes who have become Olympic athletes, Oscar Pistorius' body is a human and high-tech hybrid; the question of his status must therefore be raised. As a technological hybrid, his participation on the 'disabled sportsmen' circuit cannot be questioned. He falls within the T43 category for 'bilateral tibial amputation' and runs in the T44 category for 'single tibial amputation' where all runners use a carbon fibre blade.[13] In this category, having prostheses is therefore the norm. On the other hand, the 'ordinary' sports world, that guarantees the theoretical equality of competition, hesitates to recognise him as a legitimate sportsman within this order, because he uses a technical device. However, this reluctance, or even resistance, to authorise the integration in sports competitions of new technologies that allow athletes to optimise their performances may seem surprising. Georges Vigarello has shown how the discovery of new materials, and in particular fibres that make poles and boards more flexible, or composite structures that strengthen rackets and skis, have revolutionised motor skills in sports, generating, as he puts it, 'audacious new motor skills for new materials' (Vigarello, 1988, p. 68). The running blades used by amputated runners are part of this analysis. Banning new materials therefore seems relatively nonsensical when looked at solely from the perspective of the history of sports techniques.

However, Vigarello draws our attention to the specific case of a 'utensil' that he qualifies as 'close to cheating': a shoe sole 3–4 cm thick, known as a 'built-up sole', tested by Yuri Stepanov in 1957 during a high jump, and with which he beat the world record. The tool spread rapidly among other athletes, until in 1958 the IAAF introduced a regulation limiting sole thickness to 12.7 mm. A rapid institutional consensus thus seems to have been reached regarding the illegitimacy of such a technical 'invention' in sport, without clearly explaining in what way it differed from legitimate performance-optimising techniques.

As a mirror for the Pistorius case, the Stepanov story is nevertheless instructive. He was presented as using a 'miniature portable trampoline' to produce such a performance, so the structural link with the Pistorius controversy is a striking one. They both 'wear' their devices, which thus become in some way a very part of them as sportsmen. Indeed, it is this 'incorporation' of the technological device which is debated or refused by the sports authorities. Technical aids are here seen as artificial transformations of a sportsperson's biological identity,[14] assimilated to 'real doping' (de Léséleuc and Marcellini, 2005) and rejected as an attack on the ideal of fair competition between the 'biological pureness' or 'naturalness' at stake in sports competitions. It is therefore very interesting to see that at no time is the principle of prosthesis 'normalisation' brought up during the debates. Everything happens as if the sports organisation cannot apply the standard procedure of material normalisation to prostheses, which it nevertheless places in the category of 'technical aids'. This apparent paradox reveals a distinction that the sports institution makes between a technical aid that is added to a whole body (such as the built-up sole or the swimsuit) and a technical aid that 'replaces' a part of the body. In this way it confirms its interpretation of the runner Oscar Pistorius as a 'biotechnological hybrid' who, as such, is situated beyond the limits that define the legitimate participants in the common sporting game. The 'mixture' of technology and human that he has become, as an athlete, challenges official sports categories. Categories based on age, sex and weight have already been devised, so where does Pistorius fit? He becomes an 'uncategorisable' case for sports law.

This controversy surrounding the Pistorius case focuses on his prostheses which thus become 'highly' visible; conversely, it renders totally invisible the work needed to adjust to the prostheses and to make arrangements with the physical and social environment, work that underlies the execution of a 'performance in situation'. Indeed, Pistorius' performances, which equal and are deemed capable of beating those of able-bodied sportsmen, are associated with the 'qualities' of his technical aids, in this case the high-tech prostheses. The work of adjusting to technical aids, of learning to act with them and thus developing one's capacities, is hidden and even denied. The importance of practical arrangements with the environment is also ignored; the normalised aspect of the running track allows Pistorius to work on the fluidity of his running, but any change to the track could destroy the adjustment that has been achieved – for example, if it rains, Pistorius cannot hold the bends. Conversely, what is highlighted is the meeting point between human 'qualities' and the 'qualities of the prostheses', the latter in some way artificially modifying – fraudulently as far as the sporting world is concerned – the sportsperson's aptitudes, which thus become suspect.[15]

Conclusion: from the rehabilitation of the 'impaired' to the surpassing of 'natural' aptitudes

The biotechnological illusion or the return of the 'man–machine'

This comparison between studies of the subjective experience of a disabled person's relationship with technical aids and of the social attitude towards bodies with prostheses highlights the invisible nature of the process of adjusting to technical aids. Such invisibility creates the illusion that humans can be enhanced by technology, without any effort, learning or process of construction. One can see in this the return of a biologising and mechanical way of perceiving the human body, as if an individual's capacities were all 'given' by biology or by biotechnology, with a strange eviction of things psychological or social from the process of building human capacities. In this biotechnological illusion, everything takes place as if people's physical capacities were exclusively related to their physical aptitudes. This confusion is of the same order as that which assimilates people's physical impairments to their deficiency, that is, the harm inflicted on their organic being by illness or trauma. However, what distinguishes a subject's aptitudes from their initial capacities is the process of individual development, and in particular the processes of learning and adaptation. The biotechnological illusion contributes to the invisibility of this distinction because it suggests that technological aids will *ipso facto* improve capacities, thus removing any need for the subject to work on adjusting to the technical object or making arrangements with the environment. We like to think, in a mechanical shortcut, that cochlear implants will 'automatically' allow the deaf to hear, just as an amputated runner's prosthesis will 'automatically' allow them to achieve super-performances.

Pistorius thus appears to be a cyborg prototype, both literally (through his prostheses) and figuratively (in discourse, advertisements and representations).[16] The issue of enhancement acquires meaning within this context because we tend to forget the processes of adjustment, accommodation and arrangement that technical aids and the environment require, processes through which the status of the technical aid, as part of the person or as an outside object, is defined. In ordinary situations, the prosthesis that replaces a leg is used to strengthen the normality and humanity of the publicly perceived person. Pistorius shows his prostheses: he makes them visible, publicly perceptible and claims an identity as a 'human with prostheses'.[17]

The cult of performance and increasing debate about enhancements to the human body

A society which promotes infinite performance (just like infinite growth) cannot avoid producing technosciences for the enhancement of the human body. The scientific popularisation of this enterprise echoes the mental shortcuts which obliterate the complexity of the processes that lead from the use of prostheses to the development of new capacities. There is also the issue of limits: limits to rehabilitation (the 'normal' human), limits to sport (the biologically pure human), limits to the modern society of performance and competition. The social legitimacy of 'normalisation' through rehabilitation and technical aids is strong, based on a therapeutic principle and a logic of empowering people who must become 'ordinary'. The social legitimacy of surpassing one's sporting performances *via* technical aids is always contested in the name of the principle of 'biological purity' in competitive sports. The social legitimacy of enhancing humans through technical, biotechnological or biochemical aids in the daily competition of modern life is still open to question. Biological, mechanical and technological visions of the body are not restricted to the human being's physical and motor performances. Far from it.

Acknowledgements

The authors would like to thank Christopher Hinton and Simone Bateman for their help in translation.

Notes

1. This chapter is based on two research projects. The first concerns the use of wheelchairs, was carried out by Myriam Winance (2006, 2010) and involved observing wheelchair trials and interviews with wheelchair users (all of whom gave their consent to be interviewed for this project). The second, by Anne Marcellini and Éric de Léséleuc (Marcellini et al., 2010), looks at the controversy surrounding the 'Pistorius case'. It is based on an analysis of certain institutional sources, and of data from the press and from scientific literature.
2. Pendular walking/verticalisation is still used, but to different ends; in particular to allow people to become aware of their new bodies and for its benefits in terms of blood circulation and intestinal transit.
3. There is little literature on the history of the wheelchair in France.
4. J. Sanchez (1997) has shown how, in France, rehabilitation led to the issue of accessibilisation.
5. It would be useful to clarify what is meant by 'normal capacities'; indeed, we might hypothesise that the concept of 'normal capacities' has changed over the course of the 20th century, with respect to changes in the concept of 'personhood'. One of the constant issues is 'being able to work'.

6. As Stuart Blume (2010) argues, the social model has not shown much interest in technology.
7. Mr Doris – not his real name – is 55 years old, has been using a wheelchair for 30 years and lives alone in a house. All names are fictional to protect the identity of the persons interviewed.
8. There exists a diversity of models. Each model comes in different sizes. Some can also be adjusted in different ways: position of the seat in relation to the wheel axis, backs which can be set at different angles, and so on.
9. That is, one on each side.
10. This difference was frequently mentioned during the interviews. On the one hand, people point out the greater accessibility of places and services; on the other, they say that greater accessibility does not necessarily mean greater access to places and services.
11. We must nevertheless emphasise the specific situation of Neroli Fairhall, who used his bow while sitting in a wheelchair, and who was questioned about this issue.
12. See, for example, the Nike advertisement where Oscar Pistorius is shown against a black background, standing on his two racing prostheses, in a tight-fitting and futuristic suit, with a first-person text in which he defines himself as a 'thing': 'I was born without bones below the knee. I only stand 5 feet 2. But this is the body I have been given. This is my weapon... How I became the fastest thing with no legs...'.
13. This category-based grouping is organised with regard to the low number of athletes with bilateral amputations.
14. The 2009 controversy surrounding the new polyurethane swimsuits which were forbidden and then re-accepted by FINA (the International Swimming Federation) also provide an interesting case (see for example the article on unapproved swimsuits, "Combinaisons: la FINA fait trainer", *L'Équipe*, 19/05/2009).
15. Following the authorisation to compete in 'normal' sporting events, given by the Court of Arbitration for Sport (CAS) in May 2008, Oscar Pistorius took part in his country's (South Africa) Olympic selection process, but did not achieve the minimum results required. On the other hand, on July 19 2011, he qualified for the Athletics World Championships, taking part in August 2011. He won a medal for his participation in the 4 × 400 m relay.
16. S.L. Kurzman's work (2001, 2002) demonstrates this contrast between the daily use of prostheses on the one hand, based on what he calls a work of alignment (which we call adjustment) and, on the other hand, the production of cyborgs: 'I am not a cyborg simply because I wear an artificial limb. I see cyborg more as a subject position than an identity, and believe it is more descriptive of my position vis-a-vis the relationships of production, delivery, and use surrounding my prosthesis than my actual physical interface with it' (Kurzman, 2001, p. 382).
17. See note 12.

References

Avan L., Fardeau M. and Stiker H.-J. (1988) *L'homme réparé. Artifices, victoires, défis* (Paris: Gallimard).
Barton L. and Oliver M. (eds) (1997) *Disability Studies: Past, Present and Future* (Leeds: The Disability Press).

Blume S. (2010) 'Bringing technology back in' (oral presentation at the CERMES3: Paris).
Campbell J. (1997) '"Growing pains" disability politics – the journey explained and described', in Barton L. and Oliver M. (eds) *Disability Studies: Past, Present and Future* (Leeds: The Disability Press), pp. 78–89.
de Léséleuc É. and Marcellini A. (2005) 'Légitimité *versus* illégitimité du dopage chez les sportifs de haut-niveau. Comment se définissent les limites du non acceptable?', *Revue STAPS*, 26(70), 33–47.
Ewald F. (1986) *L'État providence* (Paris: Grasset).
Hirsch V. and Mathiot C. 'L'athlète – sans les jambes', *Libération* newspaper, 3/07/2007. Available at http://www.liberation.fr/grand-angle/0101106570-l-athlete-sans-les-jambes (05/07/2013).
Kurzman S.L. (2001) 'Presence and prosthesis: a response to Nelson and Wright', *Cultural Anthropology*, 16(3), 374–87.
Kurzman S.L. (2002) '"There's no language for this". Communication and alignment in contemporary prosthetics', in Ott K., Serlin D., Mihm S. (eds) *Artificial Parts, Practical Lives* (New York: New York University Press), pp. 227–46.
Lippi G. and Mattiuzzi C. (2008) 'Pistorius ineligible for the Olympic Games: the right decision', *British Journal of Sports Medicine*, 42(3), 160–1.
Marcellini A., Vidal M., Ferez S., de Léséleuc É. (2010) '"La chose la plus rapide sans jambes". Oscar Pistorius ou la mise en spectacle des frontières de l'humain', *Politix* (90), 139–65.
Oliver M. and Barnes C. (1998) *Disabled People and Social Policy: From Exclusion to Inclusion* (London and New York: Longman).
Quéré L. and Relieu M. (2001) *Modes de locomotion et inscription spatiale des inégalités. Les déplacements des personnes atteintes de handicaps visuels et moteurs dans l'espace public* (Paris: CEMS-EHESS).
Sanchez J. (1997) 'Enjeux concrets et symboliques de l'accessibilité', in Ravaud J.-F., Didier J.-P., Aussilloux C., Aymé S. (eds) *De la déficience à la réinsertion. Recherches sur les handicaps et les personnes handicapées* (Paris: Les Éditions de l'INSERM), pp. 139–46.
Stiker H.-J. (1997) *Corps infirmes et sociétés* (Paris: Dunod).
Thévenot L. (1994) 'Le régime de la familiarité. Les choses en personne', *Genèse*, 17, 72–101.
Tremblay M. (1996) 'Going back to Civvy Street: a historical account of the impact of the Everest and Jennings wheelchair for Canadian World War II veterans with spinal cord injury', *Disability & Society*, 11(2), 149–69.
Tremblay M., Campbell A. and Hudson G.L. (2005) 'When elevators were for pianos: an oral history account of the civilian experience of using wheelchairs in Canadian society. The first twenty-five years: 1945–1970', *Disability & Society*, 20(2), 103–16.
Vigarello G. (1988) *Une histoire culturelle du sport. Techniques d'hier et d'aujourd'hui* (Paris: Éd. Revue EPS-Laffont).
Vignier N., Ravaud J.-F., Winance M., Lepoutre F.-X. and Ville I. (2008) 'Demographics of wheelchair users in France: results of national community-based handicaps-incapacités-dépendance surveys', *Journal of Rehabilitation Medicine*, 40(3), 231–9.
Wilson D.J. (2009) 'And they shall walk: ideal *versus* reality in polio rehabilitation in the United States', *Asclepio. Revista de Historia de la Medicina y de la Ciencia*, LXI(1), 175–92.

Winance M. (2006) 'Trying out wheelchair. The mutual shaping of people and devices through adjustment', *Science, Technology and Human Values*, 31(1), 52–72.

Winance M. (2010) 'Mobilités en fauteuil roulant: processus d'ajustement corporel et d'arrangements pratiques avec l'espace, physique et social', *Politix* (90), 115–37.

7
Brain–Machine Interface (BMI) as a Tool for Understanding Human–Machine Cooperation

Selim Eskiizmirliler and Jérôme Goffette

Introduction

The abbreviation of the expression 'Brain Machine Interface' (BMI), represents a popular research area generating media excitement over a great number of promising applications. Even though medical applications are the most researched, the decision makers in the world of industry (including those interested in military applications), entertainment, security, and so on, are not simply observing what is happening in this field but are also investing in their own research. In fact, science fiction has often preceded scientists and engineers in imagining the technologies of the future. For example, what from the *Star Trek* television series of the 1960s can still be considered an improbable or unbelievable image, other than the presence on board of Mr Spock, a scientist who was a hybrid of the human and Vulcan species? The image of Captain Kirk talking to Dr McCoy through his mobile phone has already become a trivial, even a troubling picture of daily life. Robot factories may not be visible to everyone but they mass produce, maybe not the clever R2-D2 or C-3PO droids of the film *Star Wars*, but powerful, precise, high speed manipulators that can increase production while decreasing the necessary manpower. Talking to computer screens, like Mr Scott in *Star Trek*, though not widespread, is no longer a dream. The young scriptwriters of *Matrix* (1999) had to make Neo 'load the matrix' to make him capable of entering and exiting virtual life. In 2009, James Cameron's paraplegic Jake Sully simply needed a beautiful Avatar to enter and exit virtual life – an idea that Cameron had developed back in 1994, long before Meel Velliste and his colleagues published a video of a monkey feeding itself using a prosthetic arm on *Nature's* internet site (Velliste et al., 2008). More recently on *Nature's* web site, a woman with paraplegia could be seen using a robot arm controlled by her own mind to lift a bottle with a straw and drink from it (Hochberg et al., 2012).

Indeed, BMI is not only a dream of scientists and engineers, but ultimately a new kind of tool for human use. The birth of humankind is not only a birth to consciousness, as indicated by the denomination *Homo sapiens*, but also a birth into the technical world, as summarised by the term *Homo faber*, or by Hottois and Missa's expression 'Species Technica' (Hottois and Missa, 2002). Humankind and techniques are intertwined, and tools are part of our human world and even of our humanness. Consequently, as Sloterdijk (2000) has claimed, BMI can simply be considered as a new step along that path. However, this argument disregards some of the novelties introduced by this tool, as expressed in the classical Latin phrase *Nil novi sub sole* ('Nothing new under the sun') (*Eccl.* I, 9). BMI does renew some of our perspectives on humans and tools and we need multidisciplinary thinking to understand them.

As is clearly and conceptually suggested by the three letters of its acronym, BMI is composed of three main parts, but ordered top-down in a somewhat different order: (1) *Brain* → (2) *Interface* → (3) *Machine*. However, given the tendency to confound BMI with Artificial Intelligence (AI) as it is portrayed in literature, the movie industry and popular media products, a more descriptive definition of BMI with a brief reminder of basic biological facts seems necessary.

Animals perceive and control their body with their Central Nervous System (CNS) that contains more than 100 billion interconnected nerve cells (neurons). Communication between nerve cells takes a special form called spike trains that can be recorded and observed as electrical brain activity. Spike trains can be understood as a continuous series of bars, the frequency of which (the number of bars per second) changes depending on the parameters of movement such as direction, velocity, forces, and so on. A possible analogy for spike trains might be the well-known barcodes encoding the specifications of the products we buy in stores. The control of voluntary movements is achieved in a hierarchical manner following pathways that originate in the cerebral cortex, the upper part of the brain. In the case of higher vertebrates, including human beings and monkeys, the decision is made in the frontal cortex. Roughly speaking, this decision – always converted into spike trains – is first transferred to the premotor then to the motor cortex where the essential motor command is generated. This command is then sent *via* the spinal cord, transiting through some interneurons, to a pool of special neurons (called alpha motoneurons) whose role is to excite the target muscles responsible for moving an arm, leg and/or parts of the arm or leg.[1]

The *BMI* is an interrelated set of hardware and software tools that records *brain* activity, decodes the activity through *interfacing* techniques that control *machines* or final active devices called 'end effectors'.[2] The end effectors can be anthropomorphic devices (such as biomimetic robotic arms, forearms, hands) or devices of any other type (such as an on/off electrical switch, a wheelchair, clippers, a joystick or a cursor on a computer screen) that can be driven directly through recorded cortical signals. The BMI can thus be seen as a means of bypassing the spinal cord to establish a direct relation between the brain and an attached device, making it possible to restore lost motor functions when the spinal cord is damaged.

To explore the implications of BMI for human enhancement, we begin this chapter with a brief introduction to the short history of BMIs and their applications with emphasis on their use in medicine. The second part, in conjunction with first, will discuss the Interfacing/Cooperation paradigm of current and future BMI applications. Today we know that a monkey and even a human being can control a tool using only their cortical signals after a training period *via* a computer. So far, however, no one knows exactly how to explain this. What happens in the CNS during the training period? Does something change in the CNS? Does the BMI application create an image (or a map) for the use of end effectors? And what happens during the execution period? What happens if there is sensory feedback from the end effectors to the CNS? Is control possible in the opposite direction, that is, can a machine be used to control a human being *via* a BMI application?

These open-ended scientific questions raise many phenomenological and psychological questions concerning the transformations of body schema and of the way one feels one's body being eased, extended, enhanced, distorted or hampered. The last question – Can someone control somebody else using a BMI setup? – raises social, ethical and legal issues concerning the appropriate use of such technology. These questions will be addressed later in this chapter.

Brain–Machine Interface (BMI): some historical milestones

In attempting to establish the scientific achievements that led to the development of BMI, it is almost impossible to say which of its three main components has been the driving force in the emergence of this research area. Indeed, BMI was not a scientific revolution but a straightforward consequence of achievements in other fields.

Brain (Neuroscience)

Until the end of the 19th century, there was a longstanding, unresolved debate, dating back to Ancient Greece, as to the location of the control center of body movements. It confronted those who favoured the head (Platon, 428–347 BC; Herophilus, 330–260 BC; Hippocrates, 460–379 BC; Democrite, 469–399 BC) to those who favoured the heart (Aristotle, 384–322 BC; Galen, 129–203 AD).[3] A mere hundred years ago, the debate was finally settled as the mystery surrounding the structure of brain tissue was lifted. At the end of the 1880s, Santiago Ramon y Cajal used the so-called Golgi staining technique to demonstrate that brain tissue is composed of interconnected individual cells – a discovery for which he received in 1906, together with Camillo Golgi, the Nobel Prize in Physiology. Since then we have discovered that a neuron is composed of three main parts: the soma (the cell body of the neuron), the dendrites (inputs to the cell body from other neurons that can be numerous) and the axon (the output from the neuron to other neurons); communication between neurons occurs *via* the gaps (called synapses) connecting an axon to a dendrite or to a cell body or (very rarely) to an axon of another neuron. However, despite this victory of the head, the expression 'learn by heart' is still in daily use as a synonym of the verb 'to memorise'.

Although this breakthrough was a revolution in neuroscience, three other findings were also essential turning points in BMI history. The first was an event that founded the field of computational neuroscience. In 1952, Hodgkin and Huxley discovered that the action potential of a neuron could be calculated and simulated with a system of differential equations (Nobel Prize, 1963). The second was the introduction of a planar multielectrode array (MEA) for use in recording the spike profiles of cultured cells (Thomas et al., 1972). In 30 years, research has moved from a modest *in vitro* recording capacity using several electrodes to *in vivo* recording and stimulation capacity with arrays of hundreds of electrodes (Pine, 2006; Fisher et al., 2010). The third and final turning point was work done by Fetz and colleagues in 1975, in which they demonstrated experimentally that monkeys can learn to control the activity of a given cortical neuron (its firing rate)[4] through biofeedback indicating the target firing rate (volitional control). The importance of these events can easily be understood by reading Schmidt (1980) who claimed that voluntary motor commands could now be extracted from raw cortical neuron activity. Schmidt proposed using these signals to control a prosthetic device designed to restore motor functions in severely disabled patients. This was an early announcement of the approaching birth date of BMI.

Interface and Machines

The interface stage in BMI applications consists of the extraction (and/or estimation) of control parameters for end effectors (third stage) by decoding the information coded in neuronal signals (first stage). Mastering this intermediary stage represents without a doubt the most crucial step towards BMI applications. So far numerous bioinspired or non-bioinspired prediction/decoding techniques have been proposed and applied to various platforms. Powerful mathematical techniques that had already been applied, well before BMI applications, to other types of digital and analogue (natural) signals (in physics, audio engineering, electrical engineering, image processing, and so on) for feature extraction and for parameter estimation (Kalman, 1960) were, unsurprisingly, among the first candidates for non-bioinspired decoding algorithms. They acquired a dominant position at this early stage. As for the bioinspired techniques, the almost simultaneous appearance of the first simulation of neuron output (spikes = action potentials) in 1952 and of the neuron model itself in 1943 (McCulloch and Pitts, 1943) opened the doors to artificial intelligence (AI) and thus to new prediction and estimation techniques. These achievements were followed, several decades later, by the work of McClelland and Rumelhart (1986) on artificial neuron and learning models that made it possible, among other things, to obtain powerful results in translating neural data to motor control commands, comparable to the results obtained with purely mathematical methods.

Whether bioinspired or non-bioinspired, all decoding algorithms are implemented on digital computers, equipped with relevant electronic circuits used to drive the end effectors. For a 21st century reader who uses the computer as an everyday tool, it is certainly unnecessary to include a separate historical summary of computers and robots. One needs simply to compare a 1984 Apple Macintosh (processor Motorola 68000 at 7.83 MHz, RAM up to 512 Kb, equipped with a disk driver for 5 1/4 inch floppy disks that store a maximum of 720 Kb) with the computers we use today. To that comparison we must add the outstanding achievements in the kinematic and dynamic control of robotic arms (manipulators) with multiple degrees of freedom (DoF),[5] driven by different types of actuators[6] and other tools and/or toys serving human beings to interact with the environment.

The computer serves not only as a computing device but also as a physical interface between the biological data and the machines (end effectors). Using computers as an interface makes the association of the word 'intelligence' with the word 'computer' very tricky. It may even cause confusion concerning the role and the position of computers in BMI applications. What is meant by artificial intelligence will depend on how one defines the learning performed by the computer: whether it is a modelling and computing technique or a skill that can be transferred to the computer itself.

BMI work today and its applications

Without a doubt, BMI is one of the fastest growing areas of research and application. However, as mentioned earlier, it should not be considered a scientific revolution but a simple consequence of many other pioneering discoveries in the areas of neuroscience, computational neuroscience, artificial learning and computing techniques as well as in the development of artificial limbs (robotics). One could say that BMI began in 1999, the year in which it was experimentally demonstrated that a set of cortical neuron recordings (spike trains) could be used to directly control a robotic manipulator with one degree of freedom (Chapin et al., 1999). Since then, this technology has shown considerable potential for use in restoring sensory-motor skills of severely disabled patients (Schwartz et al., 2006), particularly subjects suffering from paralysis following spinal cord injuries or strokes, and amputees. Developing this potential has become the main driving force of the scientific world's growing interest in this technology.[7]

One well accepted classification of BMIs distinguishes them according to the way in which the neural signal is processed, that is according to the type of electrophysiological recording techniques used: (1) non-invasive (with electrodes placed on the surface of the skin) and (2) invasive systems (with micro-electrodes inserted in the cortex of the brain). Non-invasive BMI systems exploit electrical signals that represent a relatively global image of brain activity (such as in an ElectroEncephaloGram (EEG))[8], whereas invasive BMI systems use neural signals (like spike trains or Local Field Potential (LFP))[9] directly recorded in single or multiple cortical areas (primarily on motor and parietal cortices) via numerous surgically implanted individual electrodes or arrays. Given how little time has elapsed since the first experimental demonstration of each system, it is too early to promote one over the other. Each has its advantages and disadvantages. A comparative study thus seems necessary to provide an overall idea of the current and possible future outcomes of their respective use.

Non-invasive BMI systems

Techniques based on EEG detect the modulation of brain activity correlated to visual stimuli, gaze angle, voluntary intentions and cognitive states; however, they may differ according to the cortical area in which brain activity is recorded. Despite their long-term recording capacity and the important advantage of not exposing subjects to the risks of brain surgery, EEG (or even ElectroCorticoGram (ECoG) that requires relatively low risk surgery)[10] provides communication channels of limited capacity (with a typical transfer rate of 5–25 bits/s) and relatively slow cortical potentials (8–26 Hz) that are not always sufficient to control robotic arms (or hands) of multiple degrees of freedom. In addition, they are prone to interference from other bioelectrical signals (like ElectroMyoGram (EMG))[11] and suffer from limited temporal and spatial resolution due to the overlapping of electrical signals coming from multiple and different cortical areas. The subject must also often undergo a long training period. However these systems seem appropriate for applications such as cursor control, communication (computer-aided spelling system, Birbaumer et al., 1999), or wheelchair driving that do not require high resolution data. In these cases, BMI has been alternatively named Brain–Computer Interface (BCI) (Wolpaw et al., 2002, Luzheng et al., 2013). The most promising applications are found to be BCIs using Visual Evoked Potentials (VEPs)[12]: they identify the VEP when subjects look at particular items on a computer screen and can lead to practical and low cost systems improving the life quality of patients.

Invasive BMI systems

Invasive BMIs rely on the physiological properties of individual cortical and subcortical neurons (or pools of neurons) that modulate their activity in association with physical variables of motion. Mostly, these BMI platforms exploit the well-known correlation between discharges of cortical neurons and motor parameters of interest, but perform a reverse operation: they predict motor parameters from patterns of neuronal firing. As a result of the experimental efforts and theoretical studies that preceded the demonstration of the first invasive BMI application in rats (Chapin et al., 1999), in less than a decade there was a stream of scientific papers reporting on BMIs that reproduce primate arm reaching and a combination of reaching and grasping movements through the use of either computer cursors or robotic manipulators as end effectors.

There are several important differences among these invasive BMI systems. Following their usual top-down processing manner (Brain → Interface → Machine), they can be classified in three groups according to (1) types and places of recorded signals, (2) types of decoding techniques and algorithms and (3) types of operating device (end effectors). Each group of systems must deal with a set of specific critical issues.

Types and places of recorded signals

Most invasive BMIs have been tested on monkeys, turning them into the indispensable experimental model in this field. The systems differ according to the number of cortical implants (for example uni-site or multi-site recordings), the cortical location of implants (for example motor or parietal cortex, or both), the type of neural signal recorded (LFPs *versus* single-unit or multi-unit spike trains) and the size of the neural sample. However, the majority of BMIs tested in monkeys have relied on single cortical site recordings, either of local field potentials (Rickert et al., 2005) or of small samples of neurons or multi-unit spike trains (Carmena et al., 2003). Further, most of these small-sample, single-area BMIs use neural signals recorded in the primary motor cortex, except for rare examples such as BMI research that has focused on processing neural signals recorded in the posterior parietal (sensory) cortex (Musallam et al., 2004). Brain surgery problems such as tissue damage at implementation, inflammation reaction, bleeding and infection risks associated with the use of cables to connect brain implants to the external interface devices (recording systems) via extra-cranial head-stages, are common to all current invasive recording techniques. In addition, they can only provide good quality signals for several months (in some cases for several years but only in a limited number of species and examples) due to electrode encapsulation by fibrous tissue and/or cell death in the vicinity of the implanted electrodes and/or arrays. Therefore, despite recent work proposing the use of different types of electrodes (ceramic-based micro-electrodes, nanotechnology probes, or electrodes containing neurotrophic medium, and so on), biological compatibility and long-term functionality with safe and continuous wireless data transmission are among the foremost problems to be solved before considering safe, long-term clinical applications. Determining the number, the location (uni-site *versus* multi-site) and the types (identified *versus* non-identified) of neurons, as well as the types of signals (LFP *versus* spike trains) that a BMI system needs to perform a specific task, are still baffling problems. Even though most researchers believe that recording more data will mean more accuracy, more performance and more dexterity (Brown et. al., 2004), there is undeniable evidence that BMI systems using a small number of neurons could also be quite competitive (Ouanezar et al., 2011).

Types of decoding techniques and algorithms

With respect to the interfacing stage of invasive BMI systems, current incapacity in understanding how the central nervous system processes the motor and sensory signals generated by a huge neuron population is the main concern. Rate (frequency) coding, temporal coding and population coding have all been suggested, and many experiments with BMIs have been used to test their respective validity (Brown et al., 2004). Although the correlation between discharges (firing rate) of cortical neurons and kinematic and dynamic parameters of movements has been demonstrated by many experimental studies (Schwartz et al., 1988; Georgopoulos et al., 1988; Maier et al., 1993; Fetz, 2007), it has also been shown that close neighbour neurons may display highly distinct firing modulation patterns during the execution of a particular movement, and that single-neuron firing can vary substantially from one trial to the next even though the performed (or observed) movements remain visually identical. Among many others, averaging variables (like frequencies, amplitudes, latencies, and so on) across large populations of neurons seems to have become one of the rare, generally accepted techniques for reducing the variability of signals derived from single neurons. As for the most efficient decoding process needed to generate (predict) motor commands for artificial limbs from neuronal outputs, the BMI community remains highly divided. Competition between complex linear/nonlinear algorithms, Artificial Neural Network (ANN) based approaches and relatively simple linear regression models will probably continue in the near future. Linear methods that incorporate adaptive algorithms (that update the model parameters while the subject is training) have also been proposed. Despite their short history, it is nonetheless remarkable that no serious application of biomimetic control schemes (circuits) employing spiking or classical neuron models to establish a connection between cortical neurons and end effectors (robotic arms or other devices) has been proposed.

Types of operating device (end effectors)

So far only robotic arms (equipped with or without clippers having two degrees of freedom) driven by conventional electrical motors and joysticks, or by similar positioning devices, have been used as end effectors and, in most cases, the control paradigm was limited to prediction of the kinematic parameters of resulting movement (Wessberg et al., 2000; Lebedev et al., 2005; Velliste et al., 2008). However, future clinical applications may eventually rely on BMIs that predict dynamic control parameters. The prediction of electrical activity produced by muscles

(EMG) (Ouanezar et al., 2011), as well as the development of a new generation of artificial apparatuses using actuators based on new materials such as Electrically Active Polymers (EAPs), fibre-optic or nanomaterial-based artificial muscles, and so on, seem to have become hot topics and the scene of fierce competition. A powerful future application for BMIs that decode EMG is the construction of brain–muscle interfaces that directly stimulate the muscles of paralysed patients and thereby restore mobility by using the patient's own musculoskeletal apparatus (Peckham and Knutson, 2005). Such BMIs will most probably have to provide the means by which the hardware needed for amplification, transmission and processing of brain-derived control signals, and the muscle stimulators driven by these neural signals, can be entirely encased in the patient's body (neuroprosthetic).

The use of MEA technologies was not limited to recording neural activity. As of the late 1990s, a technique known as Deep Brain Stimulation (DBS) was successfully used to stimulate particular brain areas with the aim of altering their 'abnormal' functioning responsible for particular diseases. Applications of DBS are not considered as BMI applications but can easily be seen as Machine–Brain Interface (MBI) applications. Reducing seizures in epileptic patients by electrical stimulation of the anterior nucleus of thalamus (Fisher et al., 2010), or restoring lost motor control abilities (such as starting-stopping movements in Parkinsonian patients) by stimulating the thalamus or subthalamic nucleus (Welter et al., 2002) or the globus pallidus, have been among the most studied and promising cases. So far, DBS has been applied only in patients who have responded poorly to drugs and/or who have had severe side effects from medication. A recent publication has reported the use of MEA in restoring memory in rats (Berger et al., 2011). In this research, the brain implant is placed between two slivers of tissue (called CA1 and CA3) of the hippocampus known to be responsible for forming new memories in rats, as it is in humans. The experiment starts with the transmission of all exchanges between these two regions of the brain to a computer, while the animals are learning the rule to be rewarded. The MEA is then switched off. The paper reports that the animals forgot the rule they had learned when the researchers used a drug to shut down the activity of CA1 cells, but were able to remember the lost rule when the implanted array was switched on to replay – like a music box – the previously recorded signals. 'Flip the switch on, and the rats remember. Flip it off, and the rats forget', was Berger's comment. Nicolelis (2011) was certainly right when he used the symphony as a metaphor (cerebral symphony) to explain the neural activity of the central nervous system.

The top-down, three-stage scheme of current BMI applications, to which we have remained faithful so far, is likely to change soon. It will take the form of a closed loop by the inclusion of sensory feedbacks from the end effectors (the third stage) to the brain (the first stage). The reasons for this radical (even revolutionary) step are twofold. On the one hand, the sensory capacity of BMI devices – particularly the end effectors – may be refined to the point that they can be accepted by patients as a replacement of the subjects' own limbs. On the other hand, the image of the replaced limb (in the case of paralysed patients) or of the third arm (in the case of experimental studies with monkeys) generated by the natural plasticity of the brain, requires sensory inputs in order to be stabilised, reliable and thus constant. Current measurement and electronic technologies may easily and rapidly raise the sensing capability of today's BMI applications, presently limited to visual feedback, to a level in which proprioceptive (the joint angles), force (stiffness, exerted forces, muscle lengths) and tactile (pressure, texture, temperature) information become possible. These feedback signals could be used to train the brain to create conditions under which it undergoes experience-dependent plasticity: it would then incorporate the properties of the artificial limb into the tuning characteristic of neurons located in cortical and subcortical areas that maintain representations of the subject's own body.

Even though they may not be considered as BMI studies, recent applications which aim to provide amputee people with the feeling of their myoelectric prosthesis fingers (Raspopovic et al., 2014) suggest that such images of the external world (internal brain representations, or the body schema concept as first discussed by Head and Holmes in 1911) might be generated, making it possible for the brain to assimilate the prosthetic limb as if it were a part of the subject's own body. Moreover, the results reported from the studies concerning brain-to-brain communication in humans (Grau et al., 2014, Rao et al., 2014) showed that the questions formulated in the introduction of this chapter as open-ended scientific questions have become an actuality.

These findings open this very young research and application area to important ethical and philosophical discussions.

Brain–Machine Interface: of tools and human beings

In very general terms, BMI thus appears as a new kind of tool. In talking about tools, it is necessary to remember that the relationship between human beings and tools is all but neutral. When the knife – or its precursor – was invented, it was not only a new object among natural objects like stones, trees and rivers, but it also opened new perspectives. The invention of the knife created the ability to kill animals, carve meat, cut materials, prepare a meal, defend oneself against beasts and to fight against other human beings. Each tool offers its own perspectives, its own combination of strengths and weaknesses, its own values. A tool is not only an object among all other objects, but an object that enables or increases the capacity to perform. 'Tool' is synonymous with 'power', and subsequently with uses and rules of good usage. Finally, tools are essential components in the transformation of societies and cultures, and each tool induces its own effects.

Tools, the body and the self are not completely separate. Phenomenological philosophy (Husserl, 1989; Merleau-Ponty, 2002) and developmental psychology (Maine de Biran, 1859; Wallon, 1934; Lhermitte, 1939), with their reflections on body ownership and the body schema (Head and Holmes, 1911; Schilder, 1935; Gallagher, 2005), emphasise, from a psychological and behavioural point of view, that a tool is often integrated into the body. The most classical example is the blind person's cane that enables them to feel the environment as if the cane were a long arm or finger. Recent studies in cognitive science have provided numerous experimental confirmations of this type of embodiment and of the varied forms of this phenomenon, from quasi-bodily embodiment (Botvinick and Cohen, 1998) to just partial integration (Costantini and Haggard, 2007). In addition, the sense of ownership can be very different according to the kind of tool that is used (de Vignemont, 2011). This will depend in particular on three components: sensory impressions, motor skills and affective involvement. For example, when one takes a spoon in hand to stir a boiling soup, sensory-motor feelings induce an incorporation of the tool, but there is no affective involvement as towards a real body part (as when one fears contact with boiling liquid), and one does not have the sense of body ownership of the spoon, because it is a tool incapable of feeling and fearing high temperature. At the other extreme, the rubber hand illusion generates a protective attitude towards the rubber hand, by an assimilation of the false hand with one's own. All tools do not involve the same embodiment and sense of ownership, both of which depend on circumstances and situations.

With this background in mind, the specific case of prostheses, that is devices that imitate a part of the body or extend it, raises questions about both embodiment and ownership. When it is used, a prosthetic leg must be felt as a part of the body. This embodiment involves what is called a phantom limb, capable of projecting itself into the prosthesis as soon as it is used. Without the prosthesis, the phantom often regresses or remains latent. Indeed, our body is not only a material body but also what is called a body schema, that is the presence of the body to the psyche. Thus, when used, a leg prosthesis is present in the body schema with its different possible facets (various sensations, motor skills, some affective investment, and modulated attention). But, when removed, the prosthesis becomes an object like all the external objects, into which one is unable to project a gesture, with which one cannot feel, and towards which one does not experience the sense of immediate protection due to the body. The more the prosthesis is a source of sensations and motor expression, the more it appears affectively invested, embodied and associated with a sense of ownership. From this perspective, the tools that function with a BMI can be situated at the top of the scale of possible tool embodiment.

BMI and bodily tools: some specific characteristics

The BMI process provides a direct motor connection between the psyche and the prosthesis. The assimilation of the prosthesis into the body schema is no longer reduced to the sensory impressions experienced through the proprioception of the tool (sensations of weight, general equilibrium of the body), to the tactile impressions transmitted through the body's skin in contact with the prosthesis, and to the visual frame perceived by the eyes. Integration also results from the mentally directed ability to move the prosthesis, as one does a natural limb. If the BMI process includes a sensory feedback device, there is also a mentally directed sensation. A BMI prosthesis of a limb may indeed appear as a biomimetic limb, situated at a higher degree of embodiment than a normal prosthesis, closer to that of natural limbs. A BMI prosthesis of an eye, although with lower visual resolution than the normal eye, is nonetheless perhaps higher on the scale of embodiment than peripheral natural body parts, such as an elbow or a collarbone, because the psyche avidly projects itself onto visual data and integrates them. By facilitating sensory and/or motor connections, BMI is probably a threshold for embodiment and a sense of ownership of tools.

From this perspective, a BMI biomimetic prosthesis could be considered as a near equivalent of the bodily part it imitates; but such a point of view seems simplistic and narrow. Since tools are very often a means of increasing an aspect of natural performance, BMI prostheses should probably be viewed as potential bodily tools higher on the scale of embodiment than natural bodily parts (although the situation may be more complicated, because there are often various uses for one tool, and consequently various levels of performance). In *Limbo*, a famous science fiction novel, Bernard Wolfe (1952) describes a world in which limbs can be advantageously replaced by prostheses: artificial legs are more powerful than normal legs, and artificial arms and hands are more vivid and precise. This novel suggests that the relation between natural and artificial parts is not simply a relation between the original and its copy: it can and probably must be considered as a relation between a given functional model and a new one that can be better in certain aspects and not in others. This paradigm shift facilitates a progressive change in body metaphors and body models: from the naturalistic, well-defined, fixed, usual and standard body part representation to the mechanical, undefined, evolving, unusual and non-standard representation. The BMI prostheses accentuate this shift. If one goes a step further, the body's standard frame could disappear and be replaced by the metamorphic body paradigm of cyborgisation.

Such changes, as one can see, disrupt the conventional opposition between biological and mechanical parts. If, from a phenomenological and behavioural perspective, we compare ideas about biological and mechanical parts as they relate to human beings, the respective qualities of these parts or objects could be characterised as follows:

Table 7.1 Comparison of the main phenomenological and behavioural characteristics

The biological (that is the human body)	The mechanical (that is a car, a phone or a mixer)
Soft to the touch	Rigid
Tepid	Cold
Curved and springy	Angled and sharp
Smells organic	Smells metallic or plastic
Pulsing	Pulseless
Impossible to separate part by part (cutting a body induces a lethal risk)	All elements can be taken apart and assembled
Possesses an intimacy because the skin hides internal organs and protects the internal medium	Is merely exteriority because all parts can be dismantled and viewed; the sum of the parts equals the whole
Matter that grows, falls ill and ages	Matter that breaks down
Is gifted with intentional and autonomous behaviour	Follows deterministic patterns

It is clear that prostheses and, above all, BMI prostheses disturb this opposition. Even if they are not tepid and pulsing, they are often springy to the touch and soft. Even if they don't smell organic, they are curved and skin-mimetic. Even if they are not prone to growth, disease and ageing, subject intentionality drives them and they express gestures. BMI prostheses belong both to the world of objects and to the world of bodily subjects. Their hybridity is not only an association between two fields but above all a combination, in the strong sense of this term, that comprises the emergence of a third field. Bodily tools and prostheses are common but powerful entries into this field, far from the fiction and fashion of cyborgs. What is at stake is a deep transformation of the intimacy experienced by human beings and a progressive change from the old hidden, intimate human body to a new intimate/'extimate' paradigm. A window is thus opening onto various new perspectives.

From compensatory to complementary and supplementary prostheses

First of all, the use of BMI prostheses as compensatory devices was and still remains a pertinent issue. If some physical incapacities cannot be cured, BMI prostheses are a powerful means of remedying the handicap these incapacities incur. One might think this is merely an interesting but marginal action that concerns only disabled persons, but with the increase in lifespan, incapacities will become more prevalent. Disability as an issue will tend to spill over its present boundaries: today no longer a marginal problem, it will ultimately become a recurring one.

Secondly, if the pressure exerted on individuals to perform increases, BMI prostheses could become a part of the enhancing or anthropotechnical process,[13] that is to say they could be used for purposes that go beyond standard medical aims, the fight against disease and disability. In this direction, the possibilities are rather impressive, as illustrated by the following list.

Specialised prostheses: the bodily integration of tools

Acting directly on the tools we use can be an interesting way of performing tasks more quickly and differently. Here, the BMI process pursues the enhancement of task performance developed by tools. Moreover, it introduces a new *modular* conception of the body and of the human being. In the same way that additional tool modules can be fitted on a power drill (for piercing, sanding, screwing, wood turning, mixing, and so on), the human body could be equipped with prosthetic modules for all the tasks one can imagine. A prefiguration of such uses can be found among disabled athletes, Oscar Pistorius or Aimee Mullins for example: each possesses a special set of leg prostheses for competition (the well-known curved leaf springs) as well as other models for ordinary life (that imitate the form of the natural leg). One human being could have a rack of prostheses from which they would be able to take the one they need for the task they have to do.

Supplementary prosthesis

We must not restrict our view of prostheses to the replacement of natural parts; we ought to think about *additional* prostheses. In the laboratory, it is already possible to equip a monkey (Velliste et al. 2008) with a third functional arm (with a two fingered hand) that uses a BMI connection directed by a small number of neurons (a sort of Shiva prosthesis, with reference to the Indian deity often represented with four arms). This kind of achievement opens the way to galleries of additional tools for performing, perceiving or interfacing.

Virtualising prostheses

The BMI process involves a stage of numerical simulation. An obvious simple development in the use of prostheses could be a *virtual* hand that is directly commanded by the agent. All BMI hand prostheses could thus be both real devices and virtual hands modelled on a screen. A virtualising prosthesis does not really interface with the brain as seen in fiction, but uses the tool as an interface to act on the virtual world directly, and thus more quickly, intuitively and with greater dexterity than with a keyboard or a joystick.

Topological prostheses

Likewise, one could imagine *relocating* prostheses in different scenarios. The agent could operate a prosthesis that is situated at a distance from their body or that is a small or micro-scale prosthesis. Conversely, they could operate a macro- or mega-prosthesis functioning on a scale larger than the human body (for example a BMI excavator or a BMI crane). They could be completely enclosed in a macro-prosthesis and make it work, for example a big exoskeleton or a mega-prosthesis of the entire body (in this case, the agent psychologically embodies the prosthesis in which they are physically embedded, *cf.* the Japanese 'mechas'). The work of Kevin Warwick (see http://www.kevinwarwick.com/) illustrates some of these prospects, which are not at all unrealistic. In such situations, the body is no longer situated in the here and now, but is relocated and multi-located in bodily space that may be almost infinite. The body is not a victim of dismemberment (the body reduced to parts without relations), but the actor of what one might call 'transmemberment' in which the body is prolonged and modified both in its physical matter and in its body schema.

Cross-human prostheses

With the BMI process, one can imagine prostheses that combine the forces or senses of more than one human being. Several human beings could act together with the same prosthesis (collaborative prostheses). A group of human beings could share the sensations transmitted by a set of prostheses worn by all members of the group (perception-sharing prostheses). There could also be devices giving access to another person's nervous motor control (motor transmission prostheses) or nervous perception (perception transmission prostheses), or to an aspect of another person's emotional state (emotion transmission prostheses), and so on (see http://homes.cs.washington.edu/~rao/brain2brain/ or Rao et al., 2014). If these prospects are not that of a fusion of human minds like a sort of telepathic fusion, they involve a process for sharing intimacies that could turn out to be either wonderful or awful.

As an example of the combined use of several types of prostheses, let us imagine a surgeon with two normal arms and two additional ones. With one hand, they hold a miniature hand capable of doing micro-sutures, and with another, they hold a cannula for sucking up blood. The same surgeon possesses a zoom camera prosthesis and a direct command on the light spots. Moreover, they have direct access to a library of surgical techniques. In the surgical unit, a robot is directly monitored by a specialist in another hospital. If necessary, the specialist can take control of the surgeon's arm to execute a specific gesture for which they have rare expertise. In the staff, an assistant directly perceives the emotional state of the surgeon – by means of an emotion-sharing prosthesis – and intervenes on their metabolism if necessary, with psychotropic drugs. Obviously this is an extreme example, in a context where another life is at stake, with the usual high technology and expert know-how. But one may apply this example to more common working activities, such as industrial or office work. In any case, it gives us an initial idea, an emerging vision of the prosthetic world and its ramifications.

Conclusion: reflections on learning rhythm, autonomy and humanness

By way of conclusion, let us consider three lines of philosophical and cultural reflection: the problem of learning rhythm, the autonomy *versus* heteronomy question, and the issue of the future of humanness.

All the devices mentioned above modify the body schema and introduce variations in the sense of body ownership. Even if the body schema is probably composed of partly connected and partly embedded modules (Goffette, 2010), integrating a prosthesis requires both updating and learning. Current experiments with BMI show rapid integration and relatively easy updating, that just take a couple of seconds. However, this first level of integration probably has to be completed by a longer learning period to develop adroitness, and by another type of learning relative to body image (including affective and cultural representations of the new body image). These considerations raise questions as to how difficult or easy it is to adapt to and integrate each particular type of prosthesis, how one acquires the habit of shifting daily and quickly between different forms of embodiment, and as to the learning rhythm needed to fully acquire successive prostheses. The adaptive capability of the brain and of the psyche should not be underestimated, but nor should the work needed to achieve adaptation. The risk of confusion and mental disorder needs to be studied. Plausibly, all human beings do not have the same adaptive capabilities; thus a possible risk might be a tendency to dissociation in persons with schizophrenia, a tendency that may also show up in normal persons (as in the case of dissociations induced by hypnosis).

Secondly, using prosthetics generates possibilities of both freedom and alienation. On the one hand, they offer powerful means to perceive and act, which tends to favour autonomy. On the other hand, the use of prostheses could generate three kinds of problems: their cost and the social inequalities affecting access, dependency on software updating and material maintenance, and the need to cope with breakdowns. As in the case of other forms of enhancement, the cost of acquisition or rental and of maintenance could create a gap between the rich and the poor; moreover, in a work setting, it might generate dependency of the employee on the employer who assumes the cost. A possible solution to this problem could be a new form of State health insurance system, but even if the State accepts the principle of cost redistribution, it would probably only assume the cost of devices selected on the basis of criteria to be established. Dependency on software updating and material maintenance introduces a new relationship between the human body, industry and service. With very deeply integrated prostheses, the human body may be partly owned by or dependent upon a firm, generating a real problem with heteronomy. The last problem, that is the unavoidable breakdown of specific body parts, involves a new type of complication in human life. Not only will one have to consult a physician in case of illness, one will also have to consult specialists in prostheses and in body–prosthesis interfaces in case of breakdown (comparable to the situation of disabled persons and their devices). Given that medical practitioners must obey a code of ethics and practice guidelines, what should the status and the professional rules of these new practitioners be? How can one protect the client from abuse, in a relationship that is far from equal?

Thirdly and lastly, it is interesting to remember what the German philosopher Hans Jonas stated at the beginning of *The Imperative of Responsibility*: the capacity human beings have to change themselves is one of the new ethical problems of our era (Jonas, 1979). According to Jonas, normally, the shape of a human being is fixed and cannot be modified. The present and future situation of human beings calls for a new responsibility, requiring what Jonas called the 'principle of responsibility', better known as the 'precautionary principle'. As they modify the human body through a new connection with the brain, BMI prostheses introduce potentially important changes for human beings, at a social, economic, cultural and anthropological level. The technique is not in itself good or bad and must therefore not be condemned. From an ethical point of view, it is important to distinguish the uses that increase autonomy and the uses that engender heteronomy. A prospective effort must be promoted to prepare judicious humanist choices for humanness to come.[14]

Acknowledgements

The authors would like to thank Dr Simone Bateman and Assoc. Prof. Gulsat Aygen for having always been present to discuss many of the ideas introduced in the article as well as for their precious help with endless proofreading.

Notes

1. This very short paragraph attempting to provide a global image of what is happening in the CNS of higher vertebrates (like human beings and monkeys) before movement is a far cry from the complexity of events which occur in reality. It is given only to introduce some basic biological motor control concepts prior to the definition of BMI. In fact neither is the neuron the only type of cell in the CNS, nor does the spike profile resemble a simple bar, even though it is represented and used as such in most current BMI applications. Moreover, in addition to the more or less direct pathway between the primary motor cortex and the alpha motoneurons, the motor command passes through many other circuits and/or cortical or sub-cortical areas including the brain

stem and cerebellar nuclei whose outputs participate in the elaboration of the final command. Curious readers are strongly encouraged to consult any basic neuroscience book or lecture notes for more detailed information.
2. In robotics literature, the term 'end effector' is often used to designate the final active device (like clippers, fingers or even multi-fingered hands) attached to the robot manipulator. In this chapter, we use this term to define all types of active devices that can be controlled directly by the brain signals in BMI applications.
3. In French neuroscientific and philosophical literature, the expressions '*céphalocentristes*' and '*cardiocentristes*' are widely used to characterise the ancient philosophers who hypothesised that either the head or the heart were the body control centres. One could also use the terms 'head-centrists' and 'heart-centrists' instead, to emphasise the concept in English.
4. Firing rate: the frequency of the spikes (number of spikes per second) emitted by a neuron.
5. The term 'degree of freedom' (DoF) refers to the parameters of motion and deformation of a body or mechanical system. It is used in different contexts, depending on the application area. Physically, the free motion of a rigid object in three-dimensional space is described as having six DoF: three for translation and three for rotation. We here use the term DoF in its robotic context that refers to the mechanical joint allowing a translational or rotational motion in one direction. Therefore manipulators of multiple degrees of freedom can also be read as multi-joint manipulators.
6. Some authors use the word 'actuator' as a generic name for all types of effectors, such as machines in BMI applications, while others use it to designate the type of the actuating device like electrical motors, artificial muscles, pneumatic or hydraulic jacks, and so on. We prefer to use the term in its latter sense.
7. For a short but comprehensive technical review, see Héliot and Carmena (2010).
8. EEG: electrical activity (voltage) produced by the brain as recorded from electrodes placed on the scalp.
9. LFP: electrical activity produced by electrical current flows on multiple synaptic sites in a neuronal tissue. LFP is recorded by placing extracellular microelectrodes at a sufficient distance from individual neurons (so that the electrical activity recorded represents a cumulative or summed effect of multiple currents in a given region).
10. ECoG: electrical activity produced by the brain as recorded from electrodes placed on the epidural or subdural surface of the cerebral cortex.
11. EMG: electrical activity produced by muscles as recorded by surface (placed on the skin) or intramuscular electrodes.
12. VEP: evoked potential is an electrical activity (voltage) produced by the nervous system as a response to presentation of a stimulus. It can be measured either as EEG or EMG signal. Here we refer to the EEG signal evoked by visual stimulation.
13. For more detail on these concepts, see Goffette, Chapter 2 of this book.
14. As ter Meulen also argues in this book, Chapter 4.

References

Birbaumer N., Ghanayim N., Hinterberger T., Iversen I., Kotchoubey B., Kübler A., Perelmouter J., Taub E. and Flor H. (1999) 'A spelling device for the paralysed', *Nature*, 398, 297–8.

Berger T.W., Hampson R.E., Song D., Goonawardena A., Marmarelis V.Z. and Deadwhyler S.A. (2011) 'A cortical neural prosthesis for restoring and enhancing memory', *Journal of Neural Engineering*, 8(046017), 1–11, doi:10.1088/1741-2560/8/4/046017.

Botvinick M. and Cohen J. (1998) 'Rubber hands "feel" touch that eyes see', *Nature*, 391, 756.

Brown E.N., Kass R.E. and Mitra P.P. (2004) 'Multiple neural spike train data analysis: state-of-the-art and futur challenges', *Nature Neuroscience*, 7(5), 456–61.

Carmena J.M., Lebedev M.A., Crist R.E., O'Doherty J.E., Santucci D.M., Dimitrov D., Patil P.G., Henriquez C.S. and Nicolelis M.A. (2003) 'Learning to control a brain–machine interface for reaching and grasping by primates', *PLoS Biology*, 1, 193–208.

Chapin J.K., Moxon K.A., Markowitz R.S. and Nicolelis M.A. (1999) 'Real-time control of a robot arm using simultaneously recorded neurons in the motor cortex', *Natural Neuroscience*, 2(7), 664–70.

Costantini M. and Haggard P. (2007) 'The rubber hand illusion: sensitivity and reference frame for body ownership', *Consciousness and Cognition*, 16, 229–40.

Fetz E.E. and Finocchio D.V. (1975) 'Correlations between activity of motor cortex cells and arm muscles during operantly conditioned response patterns', *Experimental Brain Research*, 23, 217–24.

Fetz E.E. (2007) 'Volitional control of neural activity: implication for brain–computer interfaces', *Journal of Physiology*, 579(3), 571–9.

Fisher R., Salanova V., Witt T., Worth R., Henry T., Gross R., Oommen K., Osorio I., Nazzaro J., Labar D., Kaplitt M., Sperling M., Sandok E., Neal J., Handforth A., Stern J., DeSalles A., Chung S., Shetter A, Bergen D., Bakay R., Henderson J., French J., Baltuch G., Rosenfeld W., Youkilis A., Marks W., Garcia P., Barbaro N., Fountain N., Bazil C., Goodman R., McKhann G., Krishnamurthy K.B., Papavassiliou S., Epstein C., Pollard J., Tonder L., Grebin J., Coffey R., Graves N. and the SANTE Study Group. (2010) 'Electrical stimulation of the anterior nucleus of thalamus for treatment of refractory epilepsy', *Epilepsia*, 51(5), 899–908.

Gallagher S. (2005) *How the Body Shapes the Mind* (Oxford: Oxford University Press).

Georgopoulos A.P., Kettner R.E. and Schwartz A.B. (1988) 'Primate motor cortex and free arm movements to visual targets in three-dimensional space. II. Coding of the direction of movement by a neuronal population', *Journal of Neuroscience*, 8, 2928–37.

Goffette J. (2010) 'Quelques pas théoriques vers une psychogenèse du corps', *L'Évolution Psychiatrique*, 75(3), 353–69.

Grau C., Ginhoux R., Riera A., Nguyen T.L., Chauvat H., Michel B., Amengual J.L., Pascual-Leone A. and Ruffini G. (2014) 'Conscious Brain-to-Brain Communication in Humans Using Non-Invasive Technologies', *PLoS ONE* 9(8): e105225. doi:10.1371/journal.pone.0105225

Head H. and Holmes G. (1911) 'Sensory disturbances from cerebral lesions', *Brain*, 34(2–3), 102–254.
Héliot R. and Carmena J.M. (2010) 'Brain–Machine Interfaces', *Encyclopedia of Behavioral Neuroscience*, 221–5.
Hochberg L.R., Bacher D., Jarosiewicz B., Masse N.Y., Simeral J.D., Vogel J., Haddadin S., Liu J., Sydney S.C., van der Smagt P. and Donoghue J.P. (2012) 'Reach and grasp by people with tetraplegia using a neurally controlled robotic arm', *Nature*, 485, 372–5.
Hodgkin A.L. and Huxley A.F. (1952) 'A quantitative description of membrane current and its application to conduction and excitation in nerve, *Journal of Physiology*, 117, 500–544.
Hottois G. (2002), *Species Technica* (Paris: Vrin).
Husserl E. (1989) 'Ideas pertaining to a pure phenomenology and to a phenomenological philosophy', in *Second Book: Studies in the Phenomenology of Constitution (Husserliana: Edmund Husserl Collected Works)* (Dordrecht: Kluwer Academic Publishers).
Jonas H. (1979) *The Imperative of Responsibility. In Search of an Ethics for the Technological Age*, translation (Chicago: University of Chicago Press, 1985).
Kalman, R.E. (1960) 'A new approach to linear filtering and prediction problems', *Journal of Basic Engineering*, 82(1), 35–45.
Lebedev M.A., Carmena J.M., O'Doherty J.E., Zacksenhouse M., Henriquez C.S., Principe J.C. and Nicolelis M.A. (2005) 'Cortical ensemble adaptation to represent velocity of an artificial actuator controlled by a brain–machine interface', *Journal of Neuroscience*, 25(19), 4681–93.
Lhermitte J. (1939) *L'image de notre corps* (Paris: L'Harmattan).
Luzheng Bi, Xin-An Fan and Yili Liu (2013) 'EEG-based brain controlled mobile robots: a survey', *IEEE Transactions on Human–Machine Syste*ms, 43(2), 161–176.
Maier M.A., Bennett K.M., Hepp-Reymond M.C. and Lemon R.N. (1993) 'Contribution of the monkey corticomotoneuronal system to the control of force in precision grip', *Journal of Neurophysiology*, 69(3), 772–85.
Maine de Biran P. (1859) 'Essai sur les fondements de la psychologie', in *Œuvres inédites*, 1 (Paris: Dezobry, Magdeleine).
McClelland J.L., Rumelhart D.E. and the PDP research group (1986) *Parallel Distributed Processing* (Cambridge MA: The MIT Press).
McCulloch W.S. and Pitts W.H. (1943) 'A logical calculus of the ideas immanent in nervous activity', *Bulletin of Mathematical Biophysics*, 5, 115–33.
Merleau-Ponty (2002) *Phenomenology of Perception*, 2nd edn (New York: Routledge). [First French edition: Paris: NRF, Gallimard, 1945.]
Musallam S., Corneil B.D., Greger B., Scherberger H. and Andersen R.A. (2004) 'Cognitive control signals for neural prosthetics', *Science*, 305(3681), 258–62.
Nicolelis M.A. (2011) *Beyond Boundaries: The New Neuroscience of Connecting Brains with Machines and How It Will Change our Lives* (New York: Henry Holt & Co., LLC).
Peckham P.H. and Knutson J.S., (2005) 'Functional electrical stimulation for neuromuscular applications', *Annual Review of Biomedical Engineering*, 7:327–60.
Pine J. (2006) 'A history of MEA development', *Advance in Network Electrophysiology*, I, 3–23, doi: 10.1007/0-387-25858-2_1.

Rao R.P.N., Stocco A., Bryan M., Sarma D., Youngquist T.M., Wu J. and Prat C.S. (2014) 'A Direct Brain-to-Brain Interface in Humans'. *PLoS ONE* 9(11): e111332. doi:10.1371/journal.pone.0111332

Raspopovic S., Capogrosso M., Petrini M.F., Bonizzato M., Rigosa J., Di Pino G., Carpaneto J., Controzzi M., Boretius T., Fernandez E., Granata G., Oddo C.M., Citi L., Ciancio A.L., Cipriani C., Carrozza M.C., Jensen W., Guglielmelli E., Stieglitz T., Rossini P.M. and Micera S., (2014) 'Restoring Natural Sensory Feedback in Real-Time Bidirectional Hand Prostheses', *Science Translational Medicine* 6, 222ra19.

Rickert J., Cardoso de Oliveira S., Vaadia E., Aertsen A. and Rotter S., (2005) 'Encoding of movement direction in different frequency ranges of motor cortical local field potentials', *Journal of Neuroscience*, 25, 8815–24.

Ouanezar S., Eskiizmirliler S. and Maier M. (2011) 'Asynchronous decoding of finger position and of EMG during precision grip using CM cell activity: application to robot control', *Journal of Integrative Neuroscience*, 10(4), 489–511, DOI: 10.1142/S0219635211002853.

Schilder, P. (1935) *Image and Appearance of the Human Body: Studies in the Constructive Energies of the Psyche. Psyche Monographs*, 4, 1st edn (London: Kegan Paul, Trench, Trubner & Co.).

Schmidt E.M. (1980) 'Single neuron recording from motor cortex as a possible source of signals for control of external devices', *Annals of Biomedical Engineering*, 8, 339–49.

Schwartz A.B., Kettner R.E. and Georgopoulos A.P. (1988) 'Primate motor cortex and free arm movements to visual targets in three-dimensional space. I. Relations between single cell discharge and direction of movement', *Journal of Neuroscience*, 8(8), 2913–27.

Schwartz A.B., Cui X.T., Weber D.J. and Moran D.W. (2006) 'Brain-controlled interfaces: movement restoration with neural prosthetics', *Neuron*, 52, 205–20.

Sloterdijk P. (2000) *La domestication de l'être* (Paris: Mille et une nuits).

Thomas C.A., Springer P.A., Loeb G.E., Berwald-Netter Y. and Okun L.M. (1972) 'A miniature microelectrode array to monitor the bioelectric activity of cultured cells', *Experimental Cell Research*, 74, 61–6.

Velliste M., Perel S., Spalding M.C., Whitford A.S. and Schwartz A.B. (2008) 'Cortical control of a prosthetic arm of self-feeding', *Nature*, 453, 1098–101.

de Vignemont F. (2011) 'Embodiment, ownership and disownership', *Consciousness and Cognition*, 20, 82–93.

Wallon H. (1934) *Les origines du caractère chez l'enfant*, 5th edn (Paris: PUF).

Welter M.L., Houeto J.L., Tezenas du Montcel S. et al. (2002) 'Clinical predictive factors of subthalamic stimulation in Parkinson's disease', *Brain*, 125, 575–83.

Wessberg J., Stambaugh C.R., Kralik J.D., Beck P.D., Laubach M., Chapin J.K., Kim J., Biggs S.J., Srinivasan M.A. and Nicolelis M.A. (2000) 'Real-time prediction of hand trajectory by ensembles of cortical neurons in primates', *Nature*, 408, 361–5.

Wolfe B. (1952) *Limbo* (New York: Random House).

Wolpaw J.R., Birbaumer N., McFarland D.J., Pfurtscheller G., and Vaughan T.M. (2002) 'Brain–computer interfaces for communication and control', *Clinical Neurophysiology*, 113, 767–91.

8
Doping Behaviour as an Indicator of Performance Pressure

Patrick Laure and Sylvie Allouche

The concept of doping behaviour was initially developed to better understand the way performance enhancing drugs are used, in elite sports and beyond, throughout society (Laure, 1997). If the ability to identify doping behaviour is certainly interesting in itself, especially when it is related to health or social problems, the hypothesis we advocate here is that such behaviour may itself constitute a useful indicator of performance pressure. We indeed believe that the very concept of doping behaviour and subsequent studies of it could prove highly useful to anyone whose aim is to devise better targeted public health policies and campaigns dealing with enhancing drugs, including in narrower contexts like occupational medicine. If a strong causal link can be demonstrated between performance pressure and doping behaviour, this could then lead to policies that focus on the causes rather than on the effects of doping behaviour, probably with greater effectiveness, and maybe also with more justice and compassion towards the consumers.

But whereas the notion of doping behaviour has now been utilised in various linguistic contexts (French, Spanish, German), it remains less well-known in English-speaking ones. For this reason, we begin this chapter with a general overview of the concept. First, we present the notion of doping against the general background of substance consumption, and we explain how the idea of doping behaviour was derived from it. We then suggest a definition of the concept, with the aim of providing means to recognise the phenomenon in the various situations in which it may occur. We continue with a survey of the methods employed to assess its extent in society, and present some results already available in the field of sport and outside. We finally arrive at our main point – the kind of link that exists between doping behaviour and performance pressure – which leads us to suggest a tripartite model for understanding how doping behaviour can be used as an indicator of performance pressure.

Doping: a kind of substance consumption

Knowledge of the major contexts in which the consumption of various substances occurs appears as a necessary preamble to the description and understanding of the motivations of individuals who have recourse to doping. This is especially true regarding the kind of pressure they experience and, in some cases, this contributes significantly to determining the style of consumption (product, quantity, frequency, and so on). Four main areas of substance consumption can be described in this respect: nutrition, therapy, sensation seeking and the search for performance:

- Food is the only substance[1] which is essential to one's health and, contrary to all other substances, continues to be so over time: indeed you cannot starve yourself without putting your life in danger. According to a somewhat arbitrary definition, we here call 'food' any substance (apart from water) composed of carbohydrates, lipids, proteins and other mineral nutrients, whose ingestion provides calories to the body.
- In the field of therapeutics, the substances considered are mostly drugs. They are used to treat, prevent or diagnose diseases, and also modify various physiological states. Some are available only on prescription (through a doctor, a midwife, and so on); others are available for self-medication.
- Sensation seeking concerns substances that are licit (nicotine, alcohol, and so on), illicit (cannabis or cocaine for example), or diverted from their original purposes (psychotropic drugs). The consumption motive, simply summarised (we ask the experts not to hold this against us), is the quest for thrills, mainly psychological ones. The aim is to escape reality, at least at the beginning of substance use. Thereafter, it is often addiction that perpetuates consumption.
- The final area of substance consumption, the one central to this chapter, is doping behaviour, determined by the search for physical or intellectual performance in various contexts: sport, work, studies, family life, and so on. In contrast to the previous field, the goal here is to 'cling' to reality in order to overcome obstacles, rather than to elude it.

These four areas partly overlap, which means that depending on the nature of the substance, its consumption may fit into more than one area. For example, 'nutraceuticals' are 'food' enriched in 'therapeutic' substances, morphine is used both for 'therapeutic' and 'sensation seeking' purposes, and glucocorticoids are taken to fight against both inflammation ('therapeutic' field) and fatigue ('performance' field). The determinants of substance consumption therefore constitute a fluid mosaic which changes over time and according to the requirements set by the environment, or perceived as such by individuals.

From doping to doping behaviour

But what do we mean exactly by 'doping behaviour'? To begin with, the notion of doping comes from the field of sport, where it designates the use by an athlete of a substance which can be found on the 'Prohibited List' published by the World Anti-Doping Agency (WADA).[2] But there is a problem with that definition for, although functional in the domain of repression (as it marks the limit between what is allowed and what is forbidden), it cannot be used in the field of prevention. To illustrate this point, let us imagine the following situation (Laure, 2004): you are a young athlete and I am your coach. We have just talked about doping and, because doping threatens your health and is contrary to the ideal of fair play – the twin pillars of anti-doping campaigns –, I have convinced you that you should not consume any of the products that are on the WADA Prohibited List. But, later, someone else – another coach, a parent, perhaps a teammate – might tell you something quite different: 'So, you understand, you are not allowed to use these products. But, conversely this means you are allowed to use all the products which are not on the list: iron, megadoses of vitamins, creatine, tranquillisers, anti-depressants, and such like. Since these are not on the list, using them does not count as doping!' And this person would be right, despite the fact that using these substances can be dangerous for your health, or contrary to the ethical standards of sport.

Here is another example: again you are an adolescent, but this time you are studying for a school examination. Well, why not use one of the products from the list of prohibited substances? After all, you are not an athlete....

The inadequacy of the definition of doping for preventive purposes is one of the reasons why Patrick Laure proposed in 1997 the concept of 'doping behaviour' (*'conduite dopante'* in French). The expression has since been taken up by a certain number of publications (Pillard et al., 2002; Niezborala et al., 2006; Laure, 2007; Laure, 2010; Barkoukis et al., 2011; Ouédraogo, 2011), and has also been used at a national level in prevention campaigns, medical student curricula, official texts, and so on, in France (*Ministère de la Santé, de la Jeunesse, des Sports et de la Vie Associative*, 2008) and in other countries (notions of '*Doping-Mentalität*' in Germany and of '*conductas dopantes*' in Spain).

What is doping behaviour?

Now that we have recalled the context in which the notion of doping behaviour has emerged, we can clarify what we mean precisely by this expression: by definition, doping behaviour is the use of a substance in order to improve one's performance when facing an obstacle, whether real or imagined.

When examining this definition closely, one can see that we deem the nature of the substance itself of little importance, as far as the recognition of doping behaviour is concerned. The obstacle itself can also be of various kinds: a sport competition, a school examination, a job interview for example. In fact, the absence of a precise description of the obstacle points to the idea that what is determining in doping behaviour is that a forthcoming event or situation is *perceived* as an obstacle (for example, when a goal is set by a recognised team leader, and achieving that goal is felt to be very important). Indeed, doping behaviours occur whether the obstacle at its origin is real or simply imagined: what really matters is not so much a supposedly clear distinction between objective and subjective obstacles as the fact that an obstacle, externally qualified as imaginary, can be no less correlated with a specific person's doping behaviour than an obstacle externally regarded as real.

Now, from a 'managing point of view', which is not our main perspective here, things may appear a little different. In the case that someone wants to fight against medical drug abuse in general, the real or imaginary nature of the obstacle could make a difference. It is, for example, a common observation that many a brilliant student irrationally lacks confidence when sitting an exam, to the point that his lack of confidence makes him lose most of his abilities. We will call this student Paul. But if his friend James, who has not worked the whole year, sees tomorrow's exam as an insurmountable obstacle, his lack of confidence is probably less imaginary than Paul's. Now they may both manifest the same doping behaviour by going to their doctor, Dr Mary, and asking her for a miracle pill which will help them to be at their best during the exam. If the pill is not a placebo and has potential side effects, Dr Mary might differentiate between Paul and James when prescribing the pill, on the very basis of the reality of the obstacle. This little thought experiment also shows, contrary to what we said initially, that the nature of the substance, and especially the link between levels of effectiveness and danger, becomes important when adopting the managing perspective. It might turn out, for example, that Dr Mary will prescribe a real anxiolytic to Paul and a placebo or nothing to James who, contrary to Paul, has 'real' reasons to be anxious about his performance tomorrow.

But to return to the descriptive and identificatory, even diagnostic, perspective which is at stake here, what makes things even more complex is the fact that not only must one take the user's perception into account, but also that of the people around them (parents, teammates, colleagues, and so on). It is even quite possible that such an influence plays a key role in the adoption of doping behaviours, in a manner similar to that with addictive behaviours. Finally, we must also take into account the fact that performance concerns not only sporting feats, but also more general aspects of a person's whole physical and social life. For example, having a high level of performance for certain parents could consist of being able to deal with the 15 friends that their youngest son has invited to his birthday party.

Identifying doping behaviour

Now that we have a proper characterisation of our central concept, we can see that doping in sport, as it is currently understood, mostly appears as the result of a very specific kind of doping behaviour: it concerns a small proportion of the population, a limited and specifically determined number of products and it is subject to its own – essentially restrictive – regulation. According to the World Anti-Doping Code, a substance is indeed prohibited when it meets any two of the following three criteria (World Anti-Doping Agency, 2015):

- Medical or other scientific evidence, pharmacological effect or experience, that the substance...has the potential to enhance sport performance;
- Medical or other scientific evidence, pharmacological effect or experience, that the use of the substance...represents an actual or potential health risk for the athlete;
- WADA's determination that the use of the substance...violates the spirit of sport.... (p. 30).

Set against this narrow understanding of the doping phenomenon, and due to the very specific mission of WADA, we can see that the key to the concept of doping behaviour is its direct link to the notion of behaviour itself, and we think that this significantly improves our understanding of the phenomenon. But doping behaviour, as a behaviour, might sometimes be quite difficult to grasp: we must then determine how best to identify and measure it. This question is important in order to estimate, for example, the prevalence[3] of such behaviours in the general population or in specific contexts (a city, a firm, a sporting club, a university, and so on). However, unlike diseases or other kinds of behaviour, there are neither biological indicators, nor pathognomonic[4] signs or standards available to detect doping behaviours. In addition, the behaviour is sometimes deliberately hidden by the user, especially when the substances in use, such as 'street drugs', are illegal, or when they are banned in certain contexts, such as doping substances in sport. And, to this day, there is no method of reference (what we usually call a 'gold standard' in medicine) for producing nuanced and specific results; this means that the data available in the scientific literature, as illustrated below, must be utilised and compared with particular caution.

Methods for measuring doping behaviour

If we now look more closely at the various methods which provide information on doping behaviours, we find the following:

- Antidoping controls (only in sports): for over ten years, there has been an average of one to two per cent of positive analyses per year. This low result is often used to assert that doping is not a widespread problem in sport and, as a corollary, does not need to be a subject of concern.
- Participant observation: this method aims to gain a close familiarity with a given group of persons and their practices, through intensive involvement with them in their normal environment. This interesting but rarely chosen research method has, for example, been applied in the United States in the study of 17 professional body-builders (Auge and Auge, 1999). In practice, participant observation is more frequently used in the field of drug addiction.
- Face to face interviews: this technique is not frequently used (it is time consuming and the material thus obtained is difficult to analyse). It has however been applied to the study of drug use by body-builders in Scotland (Korkia and Stimson, 1997); in order to assess the way sportsmen perceive antidoping controls in the United States (Coombs

and Coombs, 1991); and to evaluate the prohibited substances most frequently used by sportsmen in Italy (Scarpino et al., 1990).
- Questionnaires: this is currently the most widely used method, as it allows for the study of a large number of subjects, in a standardised, systematic and rather inexpensive way. Since 1995, there has been an increase in questionnaires administered by telephone. Although widely used, the questionnaire method is nevertheless often criticised. On the one hand, several studies exploring the validity of answers to questions concerning personal consumption of illicit substances have concluded that the self-completed questionnaire *is* a reliable tool for gathering such information (O'Farrell et al., 2003; Secades-Villa and Fernandez-Hermida, 2003). But according to other authors, this tool creates too many biases in the answers (Lentillon-Kaestner and Ohl, 2011; Petroczi et al., 2011). As a consequence, in the last few years, different methods based on mathematical models like randomised response techniques[5] have been tried to improve the precision of this kind of study (Striegel et al., 2010).
- Analysis of biological samples (urine, blood, hair): a method that is rarely chosen (apart from antidoping controls), in particular because of its cost. It has been utilised, for example, to compare declared anabolic steroid use by weightlifters and the drugs actually detected in their urine (Ferenchick, 1996), or to quantify the use of amphetamines in amateur mountaineers in Switzerland (Roggla et al., 1993).

Studies on doping behaviour in sport

Now that we have presented the concept of doping behaviour, and the different methods available to assess it, we are in a position to provide a certain number of results, obtained through the gathering and comparison of various studies on the subject.

To begin with the field of sport, these studies have been conducted on five continents since the 1980s and were focused mainly on young athletes (Table 8.1). For a long time, the studies mainly used quantitative methods, given that their primary objective was to establish the prevalence of banned substance consumption. Later, researchers also became interested in the determinants of this consumption. The numbers of respondents in these studies (usually via questionnaires) are often high, for two main reasons. The first is that the larger the sample, the greater the details and nuances in the responses received; the second is that the margin of error is in this way reduced. From a general point of view, these investigations tend to show that doping behaviour concerns

Table 8.1 Prevalence of prohibited substance use among athletes

References	Region	Sample	Age	Substance	%
Brennan (2011)	USA	231	>18	hGH	12.0
Mattila (2010)	FIN	30,511	12–18	AAS	0.5
Striegel (2010)	GER	1,874	>18	Prohib. List	6.8
Laure (2007)	FRA	3,500	11	Prohib. List	1.2
Wanjek (2007)	GER	2,319	11–18	Prohib. List	15.1
Papadopoulos (2006)	EUR	2,650	>18	Prohib. List	2.6
Simon (2006)	GER	500	>18	Prohib. List	12.5
Beijing Univ. (2005)	CHI	4,160	12–20	Prohib. List	5.0
Laure (2003)	FRA	6,402	14–18	Prohib. List	5.0
Valois (2002)	CND	3,573	13–18	Prohib. List[6]	25.0
Laure (2001)	FRA	1,501	15–18	Prohib. List	5.0
Wichstrom (2001)	NOR	8,508	15–22	AAS	0.8
Beck (2001)	FRA	15,189	18	Prohib. List	B 5.7 G 2.5
Beck (2000)	FRA	13,957	17–19	Prohib. List	4.5

Note: AAS: androgenic anabolic steroids; hGH: human growth hormone. In this table, all data were collected through questionnaires. Note that the prevalence rates are not all comparable, as the studies did not explore the same substances.

not only adults but also young athletes, even if in the latter case, the phenomenon has remained rather marginal until now.

Doping behaviour among adults

If we now look at these studies in more detail, we can see that the majority of those conducted on adults show that the prevalence of prohibited substance use varies within a range between 5 and 15 per cent for all sporting disciplines. This percentage must however be read with caution. Indeed, not only are the substances under scrutiny sometimes different (that is, all substances on the Prohibited List, only hGH or only anabolic androgenic steroids), but populations are also disparate in terms of commitment and issues. For example, sport practitioners whose commitment is significant (more than ten hours of practice per week), and especially those for whom the stakes are high (professional and high level athletes: salary, prestige, and so on) have so far rarely agreed or have refused altogether to participate in this type of study (for example professional footballers). The various studies are therefore difficult to compare.

They all nevertheless share a common result: sporting women seem to resort to doping less often than their male counterparts. This might be the consequence of a different conception of practice, more focused on health, aesthetics or pleasure, than on confrontation, strength or virility (at least at levels where the sporting challenge stays low). But that does not necessarily mean that women are more relaxed about the way they cope with performance pressure in general. It seems reasonable to think that the kind of pressure faced by women simply entails another variety of possible responses. In order to put this gender difference into proper perspective, one would probably need to take into consideration the prevalence of recourse by women to diets and various kinds of surgery, as compared to men. But this is not something we can investigate within the limits of this chapter.

Doping behaviour among younger athletes

In the field of sports practised by younger people, three to five per cent of children and teenagers admit to having already taken a prohibited drug, at least once. It is lower in the 10–11 year old group, but the phenomenon progressively increases with age. The first doses of anabolic steroids are thus taken around 12 to 14 years, with age ranging from 7–17 for boys and 11–17 for girls.

Among young practitioners, we also observe that the prevalence of doping is higher among those who are involved in competition. And it rises in this group with the level of competition: local, regional, national or international. Indeed, as someone practising a sport climbs its hierarchy, their potential competitors come from ever wider territories, and they have to prevail over one after another. This means that they must continually achieve greater performances. Thus, the association between the increased prevalence of doping and that of the level of sporting requirement suggests that, from earliest youth, the recourse to doping behaviour is a good indicator of performance pressure.

Studies on doping behaviour outside sport

Aside from the field of sport, there have been only a few studies devoted to doping behaviour. The lack of interest in this issue, as well as the great difficulty in conducting this type of survey, partly explain why this is the case. Indeed, in the world of sport, when a study refers to doping, athletes understand what the study is about, even if they have only vague notions about what doping is or are hardly aware of the Prohibited List. On the contrary, in the corporate and educational worlds, the concept of doping behaviour does not mean much to the respondents, even when they do indeed use substances in order to increase their performance.

The studies conducted so far have then usually tested the hypothesis that substance consumption to increase performance is most common (and thus observable) in occupations which are the most vulnerable to fatigue or stress. In fact, as reported in Table 8.2, this seems to be the case in professions with high levels of 'liability' (such as managers, engineers, tradesmen, doctors, nurses, and so on), night work, work at reception counters, conveyors, and more generally wherever high output is required. Students, in particular medical students, are also concerned.

One can observe that the prevalence rates in this table vary widely according to the study considered, a phenomenon linked to the very diversity of activities involved and the types of substances used. As for the studies conducted in the world of sport, samples are most often rather large. However, according to anecdotal evidence like individual testimonies, reasons other than fatigue or stress seem to account for the adoption of doping behaviours in these broader contexts (for example, when someone wants to increase their creativity or satisfy certain social obligations). But until now, these consumptions have not been described in detail.

Table 8.2 Prevalence of doping behaviour

References	Region	Activity	Sample	Substance	%
Gay (2008)	FRA	Workers	663	All	39.2
Labat (2008)	FRA	Truck drivers	1,000	Cannabinoids Opiates Amphetamines	0.1
Lapeyre-Mestre (2004)	FRA	Workers	2,106	All	33.0
Laure (2003)	FRA	Family doctors	2,002	ANX STI	19.0 24.0
Laure (2000)	FRA	Medical students	112	All	58.0
Laure (1998)	FRA	Workers	600	All	15.0
Ponnelle (1998)	FRA	Firemen	103	ANX	8.0
OVE Survey (1997)	FRA	Students	28,141	STI	20.6
Haguenoer (1997)	FRA	Workers	1,978	All	23.6
Pidetcha (1995)	THA	Truck drivers	9,923	STI	36.6
Hughes (1992)	USA	Physicians	9,600	ANX STI	24.0 14.6

Note: All: any medicine or narcotic, ANX: anxiolytics, STI: stimulants. These results were obtained through questionnaires (most often self-administered), except in two cases – Labat and Haguenoer – where they were based on urinalyses (note the relatively large size of the two populations involved, despite the fact that biological analyses often prove more costly than questionnaire analyses).

If it is undoubtedly fruitful to compare these consumptions with those occurring in sport, an important difference still needs to be noted. Indeed, outside sport, one does not usually find the kind of hierarchy, based on the evaluation of a single set of competences, which is paramount in sports. For, whether in the general population or in more limited settings, although hierarchies usually exist, they tend to be related to a certain level of practice or responsibility. Thus, in a company, the series worker–foreman–engineer–manager does not imply that the latter performs better than the first, nor that they had to prevail against the others in order to occupy their places: each of them fulfils different useful functions and each of them is possibly the best at what they do without competing with the others (we indeed find something similar in team sports, but functions are far from being as specialised as in occupational contexts).

Similarly, one might argue that a PhD student is not 'superior' to a first year student, even though their knowledge and skills are supposed to be broader than those of first year students. But this position seems difficult to sustain because a PhD student, in mathematics or history for example, should at least be 'superior' to a first year student in the exact same sense that someone with a black belt in judo is 'superior' to someone with a white one, even if in most cases no direct confrontation is involved in the educational system. It might be the case however that, contrary to sporting contexts, competition is not necessarily harder at higher levels than at lower ones: see for example the very competitive first years of the French higher education *curricula*, whether it be the *classes préparatoires* ('preparatory classes') to the elite *Grandes Écoles*, or the first year of medical studies, that in both cases are concluded by highly competitive examinations. Notwithstanding these nuances, it appears that in the two contexts of work and study, the prevalence of doping behaviours cannot be globally related to the hierarchical positions of the individuals in the organisation. It will thus tend to be linked more generally to each person's perception of their need to gain in performance or to maintain a certain level of achievement.

Doping behaviour and performance pressure

We can now address the main point of this chapter, that is, the link between doping behaviour and performance pressure. By 'performance pressure' we mean any factor or combination of factors which increases the importance of performing a task as well as possible in a particular situation (Baumeister, 1984). In this definition, the various factors of performance pressure may be external, such as the demands of a supervisor or time constraints, but also internal, when for example someone is a perfectionist. As for the phrase 'as well as possible', it reflects the fact that the fulfilment of the task is a real challenge. For example, achieving sales targets for a salesperson is linked to the survival and development of their company, but also to their own professional promotion. In these situations, the person must prove as efficient as possible – which is often difficult.[7]

So what is the exact link between doping and performance pressure, and more specifically what is the main causal direction between the two? The field of sport, from which the very concepts of 'doping' and 'doping behaviour' come, proves once again useful in clarifying the link between the two phenomena. It is indeed well established (Tanner et al., 1995; Melia et al., 1996; Yesalis et al., 1997; Laure et al., 2004; Laure and Binsinger, 2007; Wanjek et al., 2007) that doping prevalence among young athletes:

- is higher among competitors;
- increases with levels of competition;
- is higher among athletes who are the most involved in their practices (more than ten hours of training a week).

Our main hypothesis then is that such a correlation between doping behaviour and performance pressure can be extended beyond sports to any situation in which people have the opportunity to use substances, usually drugs, to cope with competition.

But once this generalisation is acknowledged, the question of the causal direction between the two remains to be solved, even if it seems more probable that performance pressure leads to doping behaviour rather than the reverse. Even if one considers that the direction of the causal relationship appears obvious, it nonetheless needs to be established if it is to become the basis of expensive public health policies or other similar applications.

What about a reverse causal relationship indeed? Could substance use (or abuse) lead to performance pressure? Maybe not if the pressure is an external factor, such as a strong demand from a coach. But the answer could be different in the case of self-applied performance pressure. If the user perceives the substance as very helpful and effective in achieving their objective, they might therefore think: 'because I have taken the drug, now I have no choice: I *have to* be successful'. But one might answer that in such a case, there needs to be a previous environment of performance pressure (see the 'objective' we have presupposed), to which the person, let us call her Alice, was not sensitive, maybe because she felt completely 'out of the race' (we see once again how useful the sporting analogy is to understand the dynamics of performance pressure in general). But now that Alice has taken the doping substance, she feels she is 'in the race', because she has some hope of succeeding in attaining the goals the doping substance allows her to set. The race did not however suddenly materialise: it was already going on, and the drug simply led Alice to feel concerned by it, and more generally by the atmosphere of pressure to perform that was already surrounding her. But if Alice happens to take a doping substance, let us say some erythropoietin (EPO),[8] even though her environment does not require the kind of performance made possible by it, theoretically she would have no reason (apart from the placebo effect, which must not be underestimated in such sensitive contexts) to think what we had initially imagined ('Now I have no choice: I *have to* be successful'). If our reasoning here is sound, we must then recognise that a pressure to perform in a certain direction needs to be there first, even when the causal direction can be described as doping behaviour inducing performance pressure for a given context. In other words, in this interpretation, the performance pressure remains the prime mover, even when the effect of the doping substance is indeed to render its consumer sensitive to, and even maybe make them suffer from, a pressure about which, thus far, they had felt no concern.[9]

Doping behaviour and the balance of stakes and resources

Now, Lazarus and Folkman's comprehensive model of stress and coping can shed some light on the exact nature of the link between performance pressure and doping behaviour. It indeed suggests that stress can be thought of as resulting from an 'imbalance between demands and resources', or as occurring when 'pressure exceeds one's perceived ability to cope' (Lazarus and Folkman, 1984). In practice, when facing a problematic situation, one that could be considered an obstacle (a school exam or a public hearing for instance), a person conducts two evaluations. The first concerns the perceived stakes of the situation, for example a significant benefit to obtain or a challenge to take up. The second evaluation consists of an inventory of the resources the person believes they possess in order to overcome the obstacle. This assessment, which contributes to producing stress, causes a given behaviour, aimed at adapting to the situation. In this context, it appears that doping behaviour is a way not only to succeed, but also to reduce or adjust the stress induced by one's perception of a situation, with the aim of being effective. In short, if this aim is not present, drug use is not a doping behaviour, but an addiction.

We should however notice that if we were to continue along these lines, the logical conclusion would possibly lead to include addiction as another sort of coping strategy for everyday life, in other words as a form of doping behaviour (see also Yesalis et al., 1997, on this). This is a serious question related to the more classical and general debate about the difference between enhancement and treatment, normality and pathology. As it is not our purpose here to enter this debate, we will simply restrict ourselves to restating the generally admitted hypothesis that *there is* a difference between the normal and the pathological, as there is one between enhancement and therapy, and therefore also a difference between doping behaviour and addictive behaviour. It is nonetheless our contention that discussing the difference between these two concepts might be an original and possibly fruitful way of entering the more general debate we dare not address here.

Now returning to Lazarus and Folkman's model, it is in fact widely used, in occupational health psychology in particular. If we were to apply it to our own question, we could imagine a person put into a given position. If the distance between their perceived expectations and their ability to cope is too great, then whenever the stakes are considered high, the person will develop various strategies, such as hard work, relaxation methods or drug use – in other words, a doping behaviour. Should we go so far as to say that hard work is indeed just another kind of doping behaviour? We can at least say that doping behaviour appears to be one coping strategy among others in response to a situation considered problematic in a certain context. The decision to have recourse to doping substances is then guided by the person's particular perception of the stakes and resources available, which determine together with other factors the choice of doping substances and the way they are consumed.

Tripartite model linking doping behaviour to performance pressure

It is now possible to build a tripartite explanatory model linking doping behaviour to performance pressure, which states that doping behaviour occurs when three elements come together and interact:

- The first element is the person. They are characterised by their gender, age, beliefs, education, profession, the perception they have

of their freedom of action, how they usually cope with the judgment of others, and so on;
- The second would be represented by the substance: its accessibility, its cost, its nature (dietary supplement, medication, drug, and so on).
- The third would be the environment in which the person evolves and where the problematic situation occurs: family, group of friends, workplace, association, religious or cultural community, and so on; and also the more general context: political, economic, and so on.

The adoption of doping behaviour, as a translation of performance pressure exerted on the individual, results from the interaction of these three elements. For example, we can see how the pressure to perform at school that comes from the environment (elitist public policy, efforts to prevent unemployment, inducement from parents, and so on) can favour the development of capsules advertised as improving capacities for concentration, attention and memory in younger generations. Therefore parents, worried about the future of their children and wishing to make every effort for their benefit, may become interested in providing them with this kind of substance. In so doing, they would illustrate a case of 'vicarious doping behaviour' where the one who assesses the situation is not the one who consumes the product.

Of course, each of the three elements can often individually suffice to demonstrate a pressure to perform. However, the notion of doping behaviour is a more comprehensive indicator, as it can highlight their dynamics, logic and articulation. For example, some evidence suggests that performance enhancing drugs used by men and women are of different kinds (Laure, 2000). In this respect, stimulants which are used to increase energy appear to be more frequently consumed by men, whereas women more often use minor tranquilisers or dietary supplements. These practices could then constitute an objective indicator that the pressure to perform needs to be analysed differently according to gender. Along the same lines, we could also focus on the differences between the substances used in the business world (stimulants, anxiolytics) and those used in sport (peptide hormones such as hGH or EPO, anabolic androgenic steroids), which suggest differences in the performances sought.

Conclusion

Having considered these various facts and arguments, we hope it is now clear that properly assessed doping behaviours can provide precious indicators of performance pressure in a wide range of contexts: the higher the prevalence rate of a doping behaviour in a social group, the higher the performance pressure, or at least the perceived one. The development of such indicators would then have many practical applications: for example, in business contexts, managers who wish to reduce occupational stress or fatigue would be able to take performance pressure into account by measuring doping behaviours among executives, sales representatives, and so on.

But there are still many questions which need to be resolved. We could for example look more precisely into incidence rates[10] of doping behaviours. A high incidence rate might indeed indicate an increasing performance pressure in any particular context. Another line of investigation would consist of examining the ways in which performance pressure influences doping behaviours: for example, how does an increase in performance pressure affect doping behaviour? Does it lead to a higher consumption of a certain drug? In terms of dosage? Frequency? Or can it lead to a change in the substance being used (for a supposedly more effective one, like going from vitamins to amphetamines)? If such changes do occur under the influence of performance pressure, are they related to the intensity or to the various kinds of pressure that individuals experience? Obviously, many more studies are needed to improve our understanding of the association between the source, intensity, nature or repetition of a performance pressure and the doping behaviour it entails.

Notes

1. Here we understand 'substance' in the general sense of 'anything that someone ingests so that it spreads within the body (as opposed to an implant for example) and interacts with it'.
2. For more information on the WADA, see http://www.wada-ama.org/; for the 'Prohibited List', see http://www.wada-ama.org/en/World-Anti-Doping-Program/Sports-and-Anti-Doping-Organizations/International-Standards/Prohibited-List/.
3. 'Prevalence' is an indicator used to assess the occurrence of a phenomenon (a disease, a behaviour, and so on) in a defined population. It is mainly based on the statistical measure of the 'prevalence rate' that is the total number of cases existing in the population divided by the total population.
4. That is 'characteristic of a particular disease or pathologic condition'.
5. The randomised response technique is a method for asking embarrassing or even threatening questions while allowing the respondents to answer honestly and maintaining confidentiality, by creating a probabilistic relationship between the response given and the question posed.

6. Cannabis use apart from sports included.
7. And sometimes paradoxically leads to completing tasks with a lower level of performance than with less pressure (Beilock and Carr, 2001). See also the imaginary example of Paul above.
8. Erythropoietin (EPO) is a naturally occurring hormone that stimulates the production of red blood cells. EPO is used in medical practice, but also, illegally, in sports, mainly to artificially increase endurance. See also Nouvel, Chapter 5 in this book.
9. See also Goffette, Chapter 2, and Menuz, Chapter 3 in this book.
10. 'Incidence rates' are obtained by calculating the number of new cases arising in a population for a given characteristic, divided by the number of persons at risk of presenting this characteristic in a given period.

References

Auge W.K. 2nd and Auge S.M. (1999) 'Naturalistic observation of athletic drug-use patterns and behavior in professional-caliber bodybuilders', *Substance Use & Misuse*, 9, 34(2), 217–49.
Barkoukis V., Lazuras L., Tsorbatzoudis H., Rodafinos A. (2011) 'Motivational and sportspersonship profiles of elite athletes in relation to doping behaviour', *Psychology of Sport and Exercise*, 12(3), 205–12.
Baumeister R.F. (1984) 'Choking under pressure: self-conscious and paradoxical effects of incentives on skilful performance', *Journal of Personality and Social Psychology*, 46, 610–20.
Beck F., Legleye S., Peretti-Watel P. (2000) *Regards sur la fin de l'adolescence. Consommations de produits psychoactifs dans l'enquête ESCAPAD 2000* (Paris: OFDT).
Beck F., Legleye S., Peretti-Watel P. (2002) *Santé, mode de vie et usages de drogues à 18 ans. Escapad 2001* (Paris: OFDT).
Beijing Univ. (2005) Unpublished results, http://en.people.cn/200507/17/eng20050717_196544.html
Beilock S.L. and Carr T.H. (2001) 'On the fragility of skilled performance: what governs choking under pressure', *Journal of Experimental Psychology: General*, 130, 701–25.
Brennan B.P., Kanayama G., Hudson J.I., Pope H.G. Jr. (2011) 'Human growth hormone abuse in male weightlifters', *The American Journal on Addictions / American Academy of Psychiatrists in Alcoholism and Addictions*, 20(1), 9–13.
Coombs R.H. and Coombs C.J. (1991) 'The impact of drug testing on the morale and well-being of mandatory participants', *The International Journal of the Addictions*, 26(9), 981–92.
Ferenchick G.S. (1996) 'Validity of self-report in identifying anabolic steroid use among weightlifters', *Journal of General Internal Medicine*, 11(9), 554–6.
Gay V., Houdoyer E., Rouzaud G. (2008) 'Conduites dopantes en milieu professionnel. Etude sur un échantillon de travailleurs parisiens', *Thérapie*, 63, 453–62.
Grignon C. (2000) *Les conditions de vie des étudiants : enquête OVE* (Paris: Presses Universitaires de France).

Haguenoer J.M., Hannothiaux M.H., Lahaye-Roussel M.C., Fontaine B., Legrand P.M. (1997) 'Prévalence des comportements toxicophiles en milieu professionnel: une étude dans la région Nord-Pas de Calais', *Bulletin de l'Ordre des Médecins*, 80, 11–5.

Hughes P.H., Brandenburg N., Baldwin D.C., Storr C.L., Williams K.M. (1992) 'Prevalence of substance use among US physicians', *Journal of the American Medical Association*,17, 2333–9.

Korkia P. and Stimson G.V. (1997) 'Indications of prevalence, practice and effects of anabolic steroid use in Great Britain', *International Journal of Sports Medicine*, 18(7), 557–62.

Labat L., Fontaine B., Delzenne C., Doublet A., Marek M.C., Tellier D., Tonneau M., Lhermitte M., Frimat P. (2008) 'Prevalence of psychoactive substances in truck drivers in the Nord-Pas-de-Calais region (France)', *Forensic Science International*, 174, 90–4.

Lapeyre-Mestre M., Sulem P., Niezborala M., Ngoundo-Mbongue T.B., Briand-Vincens D., Jansou P., Bancarel Y., Chastan E., Montastruc J.L. (2004) 'Conduite dopante en milieu professionnel : étude auprès d'un échantillon de 2106 travailleurs de la région toulousaine', *Thérapie*, 59, 615–23.

Laure P. (1998) 'Enquête sur les usagers de l'automédication : de la maladie à la performance', *Thérapie*, 53, 127–35.

Laure P. (1997) *Les gélules de la performance* (Paris: Ellipses).

Laure P. (2000) *Dopage et société* (Paris: Ellipses).

Laure P. (2004) 'The prevention of doping in sport: protecting adolescent athletes', in Hoberman J.M. and Møller V. (eds) *Doping and Public Policy* (Odense: University Press of Southern Denmark), pp. 125–32.

Laure P. (2007) 'Dopage et conduites dopantes. Ou la quête du Graal des années 1950 à nos jours', in Tétart P. and Milza P. (eds) *Histoire du sport en France. De la Libération à nos jours* (Paris: Musée national du sport, Vuibert), pp. 275–86.

Laure P. (2010) 'Die Prävention von Dopingmentalität: der Weg über dir Erziehung', in Dannenmann F., Meutgens R., Singler A. (eds) *Sportpädagogik als humanistische Herausforderung* (Aachen: Shaker-Verlag), pp. 275–88.

Laure P. and Binsinger C. (2007) 'Doping prevalence among preadolescent athletes: a 4-year follow-up', *British Journal of Sports Medicine*, 41(10), 660–3.

Laure P., Lecerf T., Friser A., Binsinger C. (2004) 'Drugs, recreational drug use and attitudes towards doping of high school athletes', *International Journal of Sports Medicine*, 25(2), 133–8.

Lazarus R.S. and Folkman S. (1984) *Stress, Appraisal, and Coping* (New York: Springer Publishing Company).

Lentillon-Kaestner V. and Ohl F. (2011) 'Can we measure accurately the prevalence of doping?', *Scandinavian Journal of Medicine & Science in Sports*, 21(6), e132–42.

Mattila V.M., Rimpela A., Jormanainen V., Sahi T., Pihlajamaki H. (2010) 'Anabolic-androgenic steroid use among young Finnish males', *Scandinavian Journal of Medicine & Science in Sports*, 20(2), 330–5.

Melia P., Pipe A., Greenberg L. (1996) 'The use of anabolic-androgenic steroids by Canadian students', *Clinical Journal of Sport Medicine: Official Journal of the Canadian Academy of Sport Medicine*, 6(1), 9–14.

Ministère de la Santé, de la Jeunesse, des Sports et de la Vie Associative (2008) *Comment prévenir les conduites dopantes?* (Paris: Ministère de la Santé, de la Jeunesse, des Sports et de la Vie Associative).
Niezborala M. et al. (2006) 'Conduite dopante en milieu professionnel: étude auprès de 2 106 travailleurs de la région toulousaine', *Archives des Maladies Professionnelles et de l'Environnement*, 67(2), 215.
O'Farrell T.J., Fals-Stewart W., Murphy M. (2003) 'Concurrent validity of a brief self-report Drug Use Frequency measure', *Addictive Behaviors*, 28(2), 327–37.
Ouédraogo M., Goumbri W.B.F., Ouédraogo M., Liliou A.F., Guissou I.P. (2011) 'Conduites dopantes dans le sport au Burkina Faso: connaissances, attitudes et pratiques', *Science & Sports*, 26(1), 25–31.
Papadopoulos F.C., Skalkidis I., Parkkari J., Petridou E, 'Sport Injuries' European Union Group (2006) 'Doping use among tertiary education students in six developed countries', *European Journal of Epidemiology*, 21, 307–13.
Petroczi A., Uvacsek M., Nepusz T., Deshmukh N., Shah I., Aidman E.V., Barker J., Toth M., Naughton D.P. (2011) 'Incongruence in doping related attitudes, beliefs and opinions in the context of discordant behavioural data: in which measure do we trust?', *PLoS One*, 6(4), e18804. doi:10.1371/journal.pone.0018804
Pidetcha P., Congpuong P., Putriprawan T., Rekakanakul R., Suwanton L. (1995) 'Screening for urinary amphetamine in truck drivers and drug addicts', *Journal of the Medical Association of Thailand*, 78, 554–8.
Pillard F., Grosclaude P., Navarro F., Godeau E., Rivière D. (2002) 'Pratique sportive et conduite dopante d'un échantillon représentatif des élèves de Midi-Pyrénées', *Science & Sports*, 17(1), 8–16.
Ponnelle S., Vaxevanoglou X. (1998) 'Le stress au quotidien : les sapeurs-pompiers en intervention', *Archives des Maladies Professionnelles*, 59, 190–9.
Röggla G., Röggla M., Zeiner A., Röggla H., Deusch E., Wagner A., Hibler A., Haber P., Laggner A.N. (1993) 'Amphetamine doping in leisure-time mountain climbing at a medium altitude in the Alps', *Schweizerische Zeitschrift fur Sportmedizin* 41(3), 103–5.
Scarpino V., Arrigo A., Benzi G., Garattini S., La Vecchia C., Bernardi L.R., Silvestrini G., Tuccimei G. et al. (1990) 'Evaluation of prevalence of "doping" among Italian athletes', *Lancet*, 336(8722), 1048–50.
Secades-Villa R. and Fernandez-Hermida J.R. (2003) 'The validity of self-reports in a follow-up study with drug addicts', *Addictive Behaviors*, 28(6), 1175–82.
Simon P., Striegel H., Aust F., Dietz K., Ulrich R. (2005) 'Doping in fitness sports: estimated number of unreported cases and individual probability of doping', *Addiction*, 10, 1640–4.
Striegel H., Ulrich R. and Simon P. (2010) 'Randomized response estimates for doping and illicit drug use in elite athletes', *Drug and Alcohol Dependence*, 106(2–3), 230–2.
Tanner S.M., Miller D.W. and Alongi C. (1995) 'Anabolic steroid use by adolescents: prevalence, motives, and knowledge of risks', *Clinical Journal of Sport Medicine: Official Journal of the Canadian Academy of Sport Medicine*, 5(2), 108–15.
Valois P., Buist A., Goulet C., Côté M. (2002) *La performance sans drogue. Étude de l'éthique, du dopage et de certaines habitudes de vie chez des sportifs québécois* (Québec: Secrétariat au Loisir et au Sport).
Wanjek B., Rosendal J., Strauss B.M., Gabriel H.H.W. (2007) 'Doping, drugs and drug abuse among adolescents in the State of Thuringia (Germany):

prevalence, knowledge and attitudes', *International Journal of Sports Medicine*, 28(4), 346–53.

Wichstrom L. and Pedersen W. (2001) 'Use of anabolic-androgenic steroids in adolescence: winning, looking good or being bad?', *Journal of Studies on Alcohol and Drugs*, 62, 5–13.

World Anti-Doping Agency (2009) *World Anti-Doping Code* (Montreal: World Anti-Doping Agency).

Yesalis C.E., Barzukiewicz C.K., Kopstein A.N., Bahrke M.S. (1997) 'Trends in anabolic-androgenic steroid use among adolescents', *Archives of Pediatrics & Adolescent Medicine*, 151(12), 1197–206.

Part III

Visions of the Future: Lessons from Art and Fiction

9
The Earth as Our Footstool: Visions of Human Enhancement in 19th and 20th Century Britain

Christopher Coenen

Introduction

Current visions of human enhancement are rooted in visions of the future that were developed in Great Britain in the late 19th century and in the first third of the 20th century by a number of leading scientists and science-savvy authors, such as John Desmond Bernal and Herbert George Wells. The visions are thereby rooted in a historical situation in Britain which was marked by a conflict between the defenders of Christianity, traditional society and its values on the one hand, and socialist or technocratic, post-Darwinian progressives on the other. The latter group developed a worldview with strong similarities to traditional religious worldviews, in particular with regard to their eschatological elements. This new worldview – which can be termed an ideology of extreme progress – has proved to be remarkably successful, above all in recent years, and is most often known today as 'transhumanism'. It certainly serves different functions today than in the early 20th century, and its political context has changed significantly. However, the technoscientific imaginaries, which appear to be consistent, have remained more or less the same, with the exception of some updates regarding the relevant technologies and sciences. Visions of human enhancement are still a core element of this persistent ideology of extreme progress and perform an important function within it.

Today, these visions are promoted by so-called transhumanist organisations, by a range of prominent, albeit controversial researchers – such as the artificial intelligence researcher Marvin Minsky, the roboticist Hans Moravec, the nanotechnology visionary Eric Drexler and the inventor and futurologist Ray Kurzweil – by leading figures of the IT (information technologies) industry, by some libertarian or ultraliberal bioethicists, by representatives of military research funding agencies and by other science managers and technology policy actors, in particular in the area of nanotechnology and so-called converging technologies (for an overview, focusing on human enhancement, see Coenen et al., 2009). Examples of critics of the extreme progress ideology include conservative bioethicists; a fairly large number of Christian lay activists and theologians, including Pope Benedict XVI; a number of left-wing and ecologist civil society groups; a variety of scholars in the fields of science and technology studies and technology assessment; some feminists and disability activists; and natural scientists, engineers and science managers, who believe the visions of extreme progress to be unrealistic, unscientific and potentially detrimental to the public image of science or at least of the fields of research and development they are working in.

Interestingly, not only many current promoters of human enhancement but also many of its critics focus on, or at least include, far-reaching visions of extreme progress in their discussions about human enhancement. These far-reaching visions are often deemed quasi-religious and techno-eschatological, particularly by their critics.

The visionary core of the ideology of extreme progress was aptly summarised in the first article of a transhumanist manifesto that dates from the early 2000s and reads as follows:

> 'Humanity will be radically changed by technology in the future. We foresee the feasibility of redesigning the human condition, including such parameters as the inevitability of aging, limitations on human and artificial intellects, unchosen psychology, suffering, and our confinement to the planet earth'. (see Schneider, 2009, p. 97)

Current discussions of human enhancement are manifold and certainly do not take place solely in this highly visionary context. Large parts of the debate on so-called pharmacological cognitive enhancement, for example, make no reference to this broader ideology. In general, current ideologues of extreme progress focus less on rather well-established nutritional, surgical, pharmaceutical and other traditional medical enhancements, and more on enhancements that are based on or appear to emerge from advances in brain research, new neuro- and biotechnologies, nanoscience and artificial intelligence research. It is this latter group of emerging or visionary enhancement technologies that George Khushf (2005) referred to as 'second-stage enhancements'. According to him, the new tendencies in human enhancement are conceptualised as (1) 'self-aware evolution' (direct engineering of the next stages of the processes guiding the development of life through the genetic alteration of existing living systems or the direct creation of artificial life), (2) 'human–machine hybrids' or 'humanity 2.0' (in keeping with the trend towards developing technologies that make humans stronger, faster and more agile, using increasingly seamless human–machine interfaces and directly incorporating ever smarter technologies in the form of implanted chips, neural interfaces or simply remote sensing capacities) and (3) 'medical enhancements' (refining medical tools, enabling and enhancing normal human function and making possible radically new functions, introducing capacities that humans have never had before). To this list could be added drugs which, though still largely visionary, are designed to alter basic human traits (such as empathy or aggressiveness) or fundamental features of an individual's psychological identity (such as drugs for memory erasure). It is such second-stage enhancements that one has to focus on when analysing the ideology of extreme progress and its visions of human enhancement: only second-stage enhancements would have the potential to radically change the human condition and potentially justify the characterisation of the underlying visions as a kind of techno-eschatology.

The overstraining of the rationality of the modern idea of progress

In the 1960s, a philosophical debate about secularisation theories and ideologemes took place in West Germany. We can speak here of 'ideologemes', since this debate evolved against the backdrop of the then widespread use of the concept of secularisation in political and cultural discourses. In particular, in the case of so-called anti-totalitarian arguments, Communism – as practised in the Soviet Union and its sphere of influence – and Nazism were often interpreted at that time as secularised, perverted forms of religiosity. Moreover, many academic philosophers and cultural critics tended to reduce all manner of expression of political and cultural modernity to phenomena of secularisation and to somewhat deficient versions of older, pre-Modern ideas and practices.

Hans Blumenberg – still in his forties at the time – argued forcefully against this. In short, his thesis was that this notion of secularisation implies or openly goes hand in hand with a suspicion of illegitimacy concerning modernity. In his view, the critics of modernity – taking advantage of the connotations of unlawfulness associated with the term 'secularisation' (being the compulsory expropriation of church property) – often aim to classify modernity as an aberration and some of the core features of modern thought as epigonal phenomena.

Arguably, the most important proponent of this double allegation was Karl Löwith who had maintained in *Meaning of History* (Löwith, 1949) that such modern thinkers as Auguste Comte and Karl Marx replaced 'divine providence by a belief in progress and perverted religious belief into the antireligious attempt to establish predictable laws of secular history' (Löwith, 1949, p. 192). While he characterised the 'modern overemphasis on secular history as *the* scene of man's destiny', as 'a product of our alienation from the natural theology of antiquity and from the supernatural theology of Christianity', an alienation which 'is foreign to wisdom and faith' (Löwith, 1949, p. 192), he ascribed the modern idea of progress to eschatological hopes. In his view, the 'ideal of modern science of mastering the forces of nature and the idea of progress', which allow 'the world [to be remade] in the image of man', may have exclusively emerged in the West on account of the 'belief in being created in the image of a Creator-God' and 'the hope in a future Kingdom of God' – elements of Christian thought which could possibly be said to 'have turned into the secular presumption that we have to transform the world into a better world in the image of man' (Löwith, 1949, p. 203).

Blumenberg, in contrast, argued that the idea of progress developed independently from eschatology (as a consequence of several advances in various fields of human endeavour, particularly technology and science) and was only later 'drawn into the function of consciousness that had been performed by the framework of the salvation story' (Blumenberg, 1966, p. 49). Accordingly, 'the formation of the idea of progress and its taking the place of the historical totality that was bounded by Creation and Judgment' should be seen and analysed as two distinct events.

In Blumenberg's view, however, the 'authentic rationality' of the modern idea of progress was 'overextended' in this process. As 'an assertion about the totality of history, including the future', the idea of progress was 'removed from its empirical foundation' and 'forced to perform a function that was originally defined by a system that is alien to it'. It is therefore 'not so much the modern age's pretension to total competence as its obligation to possess such competence that might be described as a product of secularization' (Blumenberg, 1966, p. 66). In the process of secularisation, the 'idea of progress is driven to a level of generality that overextends its original, regionally circumscribed and objectively limited range as an assertion' (Blumenberg, 1966, p. 49).

Secularisation has again become a much-debated and analysed topic in recent decades. Marcel Gauchet (1985) and Charles Taylor (2007), to name but two of the most eminent scholars engaged in this field of research, have offered their own narratives of secularisation and accounts of the modern age. Gauchet tends to interpret the modern age as merely the latest manifestation and possibly the culmination of an age-old process in which the religious, due to its own structure of meaning, is progressively called into question. In his view, we are now even witnessing the collapse of what he calls the vestigial form of the religious, namely the progressive political ideologies, both in their reformist and revolutionary variants. While Gauchet, an atheist, believes we are about to enter an utterly post-religious age, Charles Taylor, a Catholic Christian, offers a different analysis. In his view, secularisation is the product of a 'mutation' of Latin Christendom which ushered in the modern age (Taylor, 2007). This mutation consisted of a massive reshaping of the Christian way of living and of the social order of Christian countries, particularly during and after the Reformation. The attempt to devalue or even wipe out the remnants of pagan traditions within Christianity and to establish purely Christian ways of life and genuinely Christian societies paved the way for the process of secularisation. However, our secular age should in Taylor's view be understood neither as being devoid of the religious nor as anti-Christian. Quite the contrary: the modern age appears in many aspects to be a realisation of medieval and early modern Christian yearnings for a more Christian world, embodied in such institutions and ideas as the free market, the public space and popular sovereignty. Such institutions and ideas, in Taylor's opinion, better reflect the true traditional spirit of Christianity than their feudal, aristocratic forerunners.

These authors have given us impressive and deeply insightful accounts of the genesis and problems of modernity. Blumenberg's approach, however, appears best suited to analysing early visions of human enhancement. These visions were only one element of a set of far-reaching visions concerning science, technology and the future of the human race, providing answers to eschatological questions. The notion of an overstraining of modern progressive rationality involves analysing the ideology of extreme progress to the point where a genuinely modern stance towards science, society and the future becomes one that is techno-eschatological and quasi-religious. What is more, relying on Blumenberg's account allows one to take seriously the self-perception and claims of the numerous atheist or explicitly irreligious ideologues of extreme progress – and not to foreclose the analysis by starting with a concept of secularisation that deems modern thought in general as merely derivative and often quasi-religious. Finally, opposition to modernity by traditional religion and other conservative forces is not neglected as insignificant from the outset.

The following will explore visions of human enhancement which were developed or popularised by several eminent British writers and scientists between roughly 1870 and 1930. The broader visionary endeavours of these early promoters of human enhancement, which were based on hopes for infinite progress, will also be sketched with a view to facilitating a critical analysis of the visions in question.

Early visions of human enhancement

In the course of the 19th century, gradualist geology, Darwinism and cosmology widened the time horizons of modernity in both directions. The distant past and the far future increasingly became subjects of scientific inquiry and speculation. Visions of biologically improving the human species (including, for example, a massively extended lifespan or higher intelligence) certainly date back to well before the 19th century. Recent debates have often referred, somewhat inaccurately, to Francis Bacon and the Marquis de Condorcet when arguing for a historical continuity of such visions. Be that as it may, it was not until the influence of Charles Darwin's works was widely felt that visions of a transhuman or posthuman age based on advanced science and technologies could evolve.

In the last paragraph of *The Descent of Man, and Selection in Relation to Sex* (Darwin, 1871), Darwin wrote that humanity, having risen to 'the very summit of the organic scale', may have 'hope for a still higher destiny in the distant future' (Darwin, 1871, p. 405), adding in the last sentence that '[w]e must, however, acknowledge, as it seems to me, that man with all his noble qualities, with sympathy which feels for the most debased, with benevolence which extends not only to other men but to the humblest living creature, with his god-like intellect which has penetrated into the movements and constitution of the solar system – with all these exalted powers – Man still bears in his bodily frame the indelible stamp of lowly origin' (Darwin, 1871, p. 405). At the end of the same year, Edgar Bulwer-Lytton's novel *The Coming Race* was published in which a superhuman race has slowly evolved, separate from the rest of humanity, out of a part of primordial humanity. In the 1860s and 1870s, Samuel Butler, author of the satirical utopian novel *Erewhon* (Butler, 1872), ironically wrote about an evolution of self-conscious and intelligent machines, including the sarcastic vision of human beings kept like pets by intelligent machines (plagiarised, for example, by Marvin Minsky) and anticipating what was discussed – though once again not until the second half of the 20th century – as the 'Promethean shame' (see Anders, 1956, pp. 21–95) of humans in the face of perfect machines. His notion that we are ourselves creating our own successors is today an important, albeit controversial element of the ideology of extreme progress (such as in the case of Moravec's and Minsky's visions of our future 'mind children'). As Brian Stableford has pointed out, 'speculative writers began to catch glimpses of the wealth of possibilities opened up by scientific progress' in the 1870s and 'were forced to pay far more attention to the likely impact of highly versatile new technologies on everyday life and social relations' (Stableford, 1985, p. 24). On the other hand, 'many writers began to wonder whether the idea of progress was really worth very much, and whether in fact the world might just as easily get worse as better' (Stableford, 1985, p. 24).

It was in this intellectual atmosphere that Winwood Reade – an explorer of Africa and a freethinker who had fairly close contact with Charles Darwin and who called himself his disciple – published a popular world history in 1872. It was named *The Martyrdom of Man* and influenced such diverse figures as Cecil Rhodes, Winston Churchill, Arthur Conan Doyle, Herbert George Wells, Clive Staples Lewis and George Orwell. His book ended with two chapters, 'The future of human race' and 'The religion of reason and love'. In these, Reade developed a vision of the future which paved the way for further development of the ideology of extreme progress in the 20th century. In his view, it is not by way of political reforms that better conditions for humankind can be created, but 'it is Science alone which can ameliorate the condition of the human race' (Reade, 1872, p. 519). In his vision of the far future, virtuous men, endowed with new, technoscientifically produced bodies, will colonise outer space. First, 'the whole world will be united by the same sentiment which united the primeval clan, and which made its members think, feel, and act as one. Men will look upon this star as their fatherland; its progress will be their ambition; the gratitude of others their reward. These bodies which now we wear belong to the lower animals; our minds have already outgrown them; already we look upon them with contempt. A time will come when Science will transform them by means which we cannot conjecture, and which, even if explained to us, we could not now understand, just as the savage cannot understand electricity, magnetism, steam. Disease will be extirpated; the causes of decay will be removed; immortality will be invented. And then, the earth being small, mankind will migrate into space, and will cross the airless Saharas which separate planet from planet, and sun from sun. The earth will become a Holy Land which will be visited by pilgrims from all the quarters of the universe. Finally, men will master the forces of Nature; they will become themselves architects of systems, manufacturers of worlds. Man then will be perfect; he will then be a creator; he will therefore be what the vulgar worship as a god. With one faith, with one desire' (Reade, 1872, p. 514), the humans of the future 'will labour together in a Sacred Cause: the extinction of disease and sin, the perfection of genius and love, the invention of immortality, the exploration of the infinite, and the conquest of creation' (Reade, 1872, p. 537).

As we can see, Reade was already aiming to overcome all the limits quoted in the above-mentioned transhumanist manifesto. He also believed that an understanding of the laws which regulate the complex phenomena of life would enable us to predict the future in the same way as we are already able to predict the movements of the planets. This expectation is of utmost importance for the further development of the ideology of extreme progress.

Reade's vision of a colonisation of outer space by virtuous transhuman beings may have lacked scientific imagination. However, a number of science-savvy authors such as H.G. Wells and Olaf Stapledon, not to mention leading scientists like John Burdon Sanderson Haldane, Julian Huxley and, in particular, John Desmond Bernal, developed in the first third of the 20th century many of those visions which continue to shape visionary debates on human enhancement and on other issues in science and technology today. Worthy of mention here are (1) neuroelectric visions of cyborgs, which at the time obviously had to be imagined without reference to computer technology, (2) ectogenesis, as popularised by Aldous Huxley's novel *Brave New World* (Huxley, 1932), (3) perfect control of emotions, (4) significant life-extension, (5) artificial life, (6) immortality of individual minds in a man–machine–symbiotic superstructure resembling an organism, (7) the conquest of outer space and (8) the saturation of the universe with earth-based intelligence, an idea also known from Ray Kurzweil (Kurzweil, 2005). Arguably, all the major concepts and core elements of current transhumanism, with the exception of certain far-reaching visions of computer technology and, possibly, nanotechnology, can thus be found in works by these authors.

Speculative essays about the future of mankind had an important influence on the development of the extreme progress ideology. Among the more prominent examples are *The Discovery of the Future*, published by Wells in 1902, Haldane's *Daedalus, or Science and the Future*, published in 1923, and Bernal's *The World, the Flesh, and the Devil. An Inquiry into the Future of the Three Enemies of the Rational Soul*, published in 1929 (for an important analysis of the interrelations of works by Wells, Haldane and Bernal, see Parrinder, 1995). The latter two essays were published in the popular *To-Day and To-Morrow* series, large parts of which have recently been reprinted in the United Kingdom, and whose relevance as a medium of science communication and whose cultural importance have been analysed in a recent research project funded by the European Union (see Saunders and Hurwitz, 2009). Although the essays by Wells, Haldane, Bernal and others like Julian Huxley had their playful aspects, they were meant as serious predictions of the future. Nevertheless, they were closely related to the new genre of scientific romance, a forerunner of modern science fiction (Stableford, 1985) whose most important representatives were Olaf Stapledon and, once again, Wells. Both played crucial roles in the development of the ideology of extreme progress. The hope underlying this ideology was for infinite progress of humanity or its offspring. In the words of Haldane: 'There is no theoretical limit to man's material progress but the subjection to complete conscious control of every atom and every quantum of radiation in the universe. There is, perhaps, no limit at all to his intellectual and spiritual progress' (Haldane, 1927b, p. 144).

In *The Discovery of the Future*, originally a lecture given at the Royal Institution in 1902, Wells argues that scientific progress will make rational prophecies possible in the future. He rhetorically asks 'what is there to stand in the way of our building up a growing body of forecast into an ordered picture of the future that will be just as certain, just as strictly science, and perhaps just as detailed as the picture that has been built up within the last hundred years to make the geological past?' (Wells, 1902, p. 57f.). Unsurprisingly, his answer is that it is the challenge of predicting the future of humanity. In his view, however, this problem will be solved by combining a post-Darwinian perspective with thoroughly scientific research on historical and social issues. Wells predicts that the 20th century will witness an even higher pace of progress, propelled by a growing, organised army of scientists. In much the same way as current radical promoters of human enhancement, Wells firmly believes in ever-accelerating technoscientific progress, while at the same time pondering over catastrophic events which may bring progress temporarily or even permanently to a halt (for an analysis of these and other tensions in the thought of Wells, see, for example, Hillegas, 1967; Parrinder, 1995, p. 29ff.; Stableford, 1985, p. 55ff.). He wrote:

> And finally, there is the reasonable certainty that this sun of ours must some day radiate itself towards extinction.... There surely man must end. That of all such nightmares is the most insistently convincing. And yet one doesn't believe it. At least I do not. And I do not believe in these things, because I have come to believe in certain other things, in the coherency and purpose in the world and in the greatness of human destiny. Worlds may freeze and perish, but I believe that there stirs something within us now that can never die again. (Wells, 1902, p. 87f.)

The first person plural in this last sentence needs some explanation, however: as Reade had done before him, and Haldane, Bernal and numerous transhumanists have done since, Wells also turns the attention of his readers to a far future in which successors of present humanity conquer outer space:

> We are creatures of the twilight. But it is out of our race and lineage that minds will spring that will reach back to us in our littleness to know us better than we know ourselves, and that will reach forward fearlessly to comprehend this future that defeats our eyes. All this world is heavy with the promise of greater things, and a day will come – one day in the unending succession of days – when beings, beings who are now latent in our thoughts and hidden in our loins, will stand upon this earth as one stands on a footstool, and laugh, and reach out their hand amidst the stars. (Wells, 1902, p. 94f.)

The image of the earth as a footstool for (post-)humanity epitomises what Reade had been writing and what had been taken up by Wells and, to some degree, by Haldane and Bernal, namely that we should revere our own power rather than the Christian God. In his Christian rationalist *Essay on Man*, Alexander Pope (1733) wrote:

> Ask for what end the heavenly bodies shine, Earth for whose use? Pride answers 'tis for mine: For me kind nature wakes her genial power, Suckles each herb, and spreads out ev'ry flower; Annual for me, the grape, the rose, renew The Juice nectarious, and the balmy dew: For me the mine a thousand treasures brings; For me health gushes from a thousand springs; Seas roll to waft me, suns to light me rise; My footstool earth, my canopy the skies.

Like Reade before him, Wells is not afraid to express that pride. While popular Christian depictions tended to soften the biblical image of the

Earth as God's footstool, this image can in fact not only be impressive but also daunting and humiliating for a modern mind. One might be driven to imagine humanity as an insect populace at the feet of a gigantic being. The image of Wells therefore appears to be an example of a misguided attempt at human 'self-assertion' in the sense of Blumenberg – misguided inasmuch as it replaces the frightening Godly power with a heroic future (post)humanity which is imagined as a kind of god-like collective.

Like our radical promoters of human enhancement today, Wells was thus already fascinated by the posthuman, as can also be seen from the following remark:

> This fact, that man is not final, is the great, unmanageable, disturbing fact that rises upon us in the scientific discovery of the future; and to my mind, at any rate, the question, What is to come *after* man? is the most persistently fascinating and the most insoluble question in the whole world. (Wells, 1902, p. 79; *italics in the original*)

While Wells – who closely cooperated with the journal *Nature* and other core institutions of the nascent modern science system and soon became a public intellectual of global renown – had, since the late 19th century, effectively been acting 'as an imaginative intermediary between the scientists and the general public' (Meadows, 2004, p. 183), things were changing in the heated and strongly utopian political atmosphere after the First World War: a small number of influential young scientists directly addressed the general public with technocratic visions of eschatological longings finally fulfilled. Inspired and heavily influenced by Wells and Haldane, it was Bernal, in his 1929 essay *The World, the Flesh & the Devil*, who elaborated on the core visions of the ideology of extreme progress with technoscientific details. Above all, what we encounter here is mechanical man, the product of an age-long merging of humans – or rather, human brains (Bernal, 1929, p. 36ff.) – with machines, a being characterised by Bernal as the new man of the future who 'must appear to those who have not contemplated him before as a strange, monstrous and inhuman creature, but...is only the logical outcome of the type of humanity that exists at present' (Bernal, 1929, p. 42). In his view, normal man is an evolutionary dead end, while mechanical man, who apparently breaks with organic evolution, is actually more in the true tradition of a further evolution. This essay also includes, amongst other things, a society based on ectogenesis, sketches of space colonies in some technical detail, artificial life, brain–machine interfaces and the vision of a human zoo of happy utopians on planet Earth which is secretly controlled by a transhuman or posthuman technoscientific elite operating in outer space (Bernal, 1929, p. 73). Bernal, an almost fanatic Catholic as a boy, opined in *The World, the Flesh & the Devil* that, when thinking of the future, even the least religious of men all retain in their minds an idea of some transcendental, superhuman event which will bring the universe to perfection or destruction (Bernal, 1929, p. 74). He wrote that all human beings want the future to be mysterious and full of supernatural power, and argued in a highly un-Marxist manner that these aspirations have built our material civilisation and will go on building it in the future, but he adds we are now 'on the point of being able to see the effects of our actions and their probable consequences in the future; we hold the future still timidly, but perceive it for the first time, as a function of our own action' (Bernal, 1929, p. 74). Bernal also proposed a kind of technical solution to realise a new kind of union of individuals, based on brain-to-brain interfaces and the dissolution of the human body in a collective techno-biological superstructure in which the individual minds, in the words of Reade, 'think, feel, and act as one'. In passing, immortality is invented as well. In the words of Bernal:

> Death would still exist...; it would merely be postponed for three hundred or perhaps a thousand years, as long as the brain cells could be persuaded to live in the most favourable environment, but not forever. But the multiple individual would be, barring cataclysmic accidents, immortal, the older component as they died being replaced by newer ones without losing the continuity of the self, the memories and feelings of the older member transferring themselves almost completely to the common stock before its death. And if this seems only a way of cheating death, we must realise that the individual brain will feel itself part of the whole in a way that completely transcends the devotion of the most fanatical adherent of a religious sect.... It would be a state of ecstasy in the literal sense, and this is the second great alteration that the compound mind makes possible. Whatever the intensity of our feeling, however much we may strive to reach beyond ourselves or into another's mind, we are always barred by the limitations of our individuality. Here at least those barriers would be down: feeling would truly communicate itself, memories would be held in common, and yet in all this, identity and continuity of individual development would not be lost. (Bernal, 1929, p. 43f.)

However, in Bernal's view, division of labour would soon set in again: 'to some minds might be delegated the task of ensuring the proper functioning of the others, some might specialise in sense reception and so on. Thus would grow up a hierarchy of minds that would be more truly a complex than a compound mind' (Bernal, 1929, p. 44). And in a way which is quite typical of the ideology of extreme progress, Bernal goes on:

> The complex minds could, with their lease of life, extend their perceptions and understanding and their actions far beyond those of the individual. Time senses could be altered: the events that moved with the slowness of geological ages would be apprehended as movement, and at the same time the most rapid vibrations of the physical world could be separated. As we have seen, sense organs would tend to be less and less attached to bodies, and the host of subsidiary, purely mechanical agents and preceptors would be capable of penetrating those regions where organic bodies cannot enter or hope to survive. The interior of the earth and the stars, the inmost cells of living things themselves, would be open to consciousness through these angels, and through these angels also the motions of stars and living things could be directed. This is perhaps far enough; beyond that the future must direct itself. Yet why should we stop until our imaginations are exhausted. Even beyond this are foreseeable possibilities. (Bernal, 1929, p. 44f.)

One of these possibilities is the creation of improved artificial life which will be used to reengineer the human brain, the only relic of the human body in the far future. Yet Bernal's controlled flight of fancy does not stop there:

> The new life would be more plastic, more directly controllable and at the same time more variable and more permanent than that produced by the triumphant opportunism of nature. Bit by bit the heritage of the direct line of mankind – the heritage of the original life emerging on the face of the world – would dwindle, and in the end disappear effectively, being preserved perhaps as some curious relic, while the new life which conserves none of the substance and all of the spirit of the old would take its place and continue its development.... Finally, consciousness itself may end or vanish in a humanity that has become completely etherealised, losing the close-knit organism, becoming masses of atoms in space communicating by radiation, and ultimately perhaps resolving itself entirely into light. (Bernal, 1929, p. 46)

Once again, Bernal stops short of defining an end to progress, as he continues: 'That may be an end or a beginning, but from here it is out of sight' (Bernal, 1929, p. 46).

It is worth mentioning that neither Bernal nor Haldane ever entirely abandoned their visions of extreme progress. Even when they had become globally prominent scientists and Communist public intellectuals – those of their manifold roles which are best remembered – they still adhered to these visions. From the 1930s onwards, both tempered their obvious contempt for a happy yet technoscientifically stagnant utopia and became heavily involved in the social and political conflicts of their times, emphasising again and again the social usefulness of science.

Their early essays, however, while belittled or idealised by their Marxist admirers, had an astounding impact on cultural discourse on science and technology. Wells, Haldane and Bernal forcefully established a new, post-Darwinian Baconianism in this discourse, devoid of Christian framings, focused on the application of science, oriented towards a far-future in outer space, and quasi-eschatological in its nature. A major literary reaction to this new atheist and futurist Baconianism was the development of the classical dystopian thought of the 20th century (see, for example, Hillegas, 1967). Haldane's influence on *Brave New World* is well-known, as is the fact that this still highly influential novel, written in the year 1931, was originally intended as a direct attack against Wellsian utopianism. Less well-known is the fact that in 1926 Charlotte Franken, Haldane's wife at the time, had published the dystopian novel *Man's World*, using the visions of her husband and anticipating *Brave New World* to a significant degree. Also well documented is the influence of Wells' scientific romances on the first great dystopian novel of the 20th century, namely Yewgeny Zhamyatin's *We*, written in the early years of the Soviet Union. Last but not least, George Orwell's polemical treatment of the Wellsian utopia in his first widely successful book, *The Road to Wigan Pier* in 1937, is testimony to the relevance of the ideology of extreme progress for the utopian and dystopian tradition. In a certain sense, the authors of the classical dystopias of the 20th century primarily reacted to the development of this ideology. Since these dystopias – *Brave New World* in particular – are still highly relevant for public and professional ethical discourses on emerging technologies, the sociotechnical imaginaries of Wells, Haldane and Bernal are still with us. The same holds true for popular Christian reactions to this ideology, namely fictional works by C.S. Lewis and his friend J.R.R. Tolkien. Lewis directly polemicised against Haldane and other ideologues of extreme progress and even wrote an essay on *The Abolition of Man*, published in 1943, which is still influential among religious and cultural conservatives, including prominent bioethicists such as Leon Kass. The science fiction of Lewis also directly confronts the ideology of extreme progress, including *ad hominem* attacks. Except in some of his letters, Tolkien did not directly polemicise against this ideology, but his immensely popular trilogy *The Lord of the Rings* can in many aspects be read as a critique of the visions of Wells, Haldane, Stapledon and Bernal (see, for example, Hogan and Clarfield, 2007).

The classical dystopian tradition and influential Christian conservative critiques of technoscientific progress are thus deeply indebted to the imagination of the ideologues of extreme progress – and both traditions are still shaping cultural discourse on science and technology today. The cultural influence of this ideology, including the radical visions of human enhancement, does not end with its impact on these two traditions, however. While only very few transhumanists are aware of the pioneering role that Reade played in the development of their worldview, the influence of Wells, Haldane, Bernal and Julian Huxley on today's transhumanism is widely recognised. More importantly, the science fiction genre, including for that matter scientific romances, was significantly shaped by Wells and strongly influenced by the essays of Haldane, Bernal, Julian Huxley and others (Stableford, 1985, p. 154ff). Stapledon had been inspired by Wells and Haldane and, in turn, inspired countless science fiction authors. Bernal's *The World, the Flesh & the Devil* has been characterised by George Slusser (2009) as the masterplot of science fiction. Other scholars as well as renowned science fiction writers also credit Haldane and Bernal with a pioneering role for science fiction. Given that science fiction is an important element of our popular culture and is closely related to technoscience itself, this influence is one reason to believe in the persistence of the ideology of extreme progress and its radical visions of human enhancement in the near future.

In a certain sense, radical visions of human enhancement are at the heart of current technoscience. In recent years, Alfred Nordmann has outlined the contours of technoscience, with a special view to nanotechnology and other so-called 'converging technologies'. In his view, there is ignorance at the heart of science that is not the kind of ignorance that can be overcome by a quest for knowledge, but 'an endemic ignorance that results from inherent difficulties in pursuing certain lines of critical questioning' (Nordmann, 2008a, p. 107). This relates, for example, to the technoscientific replacement of knowledge by know-how and to the uncertainty concerning the status of the objects of technoscientific research. This uncertainty corresponds to an uncertainty about the contents and limits of entire fields of research and development, as in the case of nanotechnology. Nordmann also places emphasis on the fundamental distinction between socially embedded or infrastructural technologies and individualised consumer technologies like the various supposed human enhancement technologies that 'require as their prerequisite a competitive world which produces intense pressures and unhappiness in individuals' (Nordmann, 2008a, p. 128). He has persistently argued for a shift of attention from our supposed historical destiny in a world transformed through individual human enhancement and other visions of extreme progress to current claims on our world of experience in the name of more plausible technoscientific visions or existing technologies such as deep brain stimulation. In his view, we need an evolutionary approach to technology assessment and other future-oriented accompanying research. Such an approach would place the future in a genuinely historical perspective. Nordmann believes that in one respect at least, the future of technical development will be like its past:

> Historical analysis shows that it was never possible to predict, let alone derive the future from the past. This will surely hold for our attempts to predict the future, too. From the point of view of the past, the future is always open. History also shows, however, that the present is always indebted to the past, that the explanations of the present lie in the past, and that the present is hardly open to arbitrary shaping. From the point of view of the present or of the future, these are to a large extent determined by the past. An evolutionary understanding would therefore consider the future undetermined but also deny that we can shape it at will. (Nordmann, 2008b, p. 45)

The ideologues of extreme progress agree that the future is undetermined, but believe that we can shape it at will and, for that reason, will progressively be able to predict it. Through their visions, Wells, Haldane and Bernal have in fact shaped their future, the time in which we now live. We can find traces of their influence in such diverse fields of scientific and scholarly practice as information science, synthetic biology, and science and technology studies.

Besides their manifold contributions to science and – as prolific science popularisers – to less speculative science communication, Wells, Haldane, Bernal and their contemporaries introduced a powerful myth into nascent technoscience. Haldane wrote in 1932 about his visions of extreme progress:

> Such speculations as these are very far from idle. They are eminently desirable, because man does not generally even know what he wants,

much less how to get it. A discussion of possibilities will have two effects. It will enable people to come to some opinions as to the possible goal of human evolution.... And it will focus attention on the necessity for more knowledge before we can even suggest means of attaining that goal. Pictures of the future are myths, but myths have a very real influence in the present. ... Our greatest living mythologist, Wells, is certainly influencing the history of the future, though probably in ways which he does not suspect. The time will probably come when men in general accept the future evolution of their species as a probable fact, just as to-day they accept the idea of social and political progress. We cannot say how this idea will affect them. We can be sure that if it is accepted it will have vast effects. It is the business of mythologists to-day to present that idea. They cannot do so without combining creative imagination and biological knowledge. (Haldane, 1932, p. 98f.)

Today, the visions of these early mythopoets of technoscience unfold their ideological potential in a world in which the idea of social and political progress has lost much ground, and the concepts of steering evolution and human enhancement have, in turn, attracted a significant number of followers within the science system and, possibly, the wider public. The recent resurgence of visions of extreme progress epitomises the devaluation of the idea of social progress in favour of an ideology which commits technoscience to the realisation of eschatological hopes.

As mentioned before, another problematic feature of the current debates on human enhancement is that the self-perception of the current ideologues of extreme progress as legitimate heirs of the age of Enlightenment and of modernity at large tends to be reaffirmed by their critics, many of whom are conservative. The Enlightenment is thus reduced to its high esteem for science and technology, while the Age of Reason is portrayed as an era of technophilia and of shallow notions of progress. If these critics refer to any Enlightenment thinkers at all, they will often use Condorcet and Julien Offray de La Mettrie as straw-men. We may thus retain the fact that prominent instances of a Christian critique of the ideology of extreme progress, such as Pope Benedict XVI's 'Second encyclical on Christian hope' (see Coenen, 2010), still exist today; these critics remain faithful to a traditional, polemically anti-modern worldview against which Reade, Wells, Haldane and Bernal argued and fought. In a certain sense, the ideology of extreme progress, including the radical 'second stage' of visions of human enhancement, can be seen as an aggressive defence of the modern idea of progress by practitioners or promoters of technoscience to counter the similarly aggressive attempts of Christian and other conservatives to delegitimise this very idea.

Conclusion: an enemy of science from within

We may conclude that we have now arrived at a position from which we can historically and ideologically locate the radical visions of human enhancement. These are a core element of an ideology of extreme progress which was developed as a means of further weakening the existing social order and its religious values. While this original function has not entirely disappeared, as the reactions of some religious figures show, the visions are now rarely combined with a radical critique of society.

This worldview is an ideology of extreme progress, in the sense that the ideologues always aspire to highlight the extremes of progress as imaginable at a given moment on the basis of a scientific worldview. In doing so, they attempt to demonstrate, ostentatiously *en passant*, that by reaching these extremes, the eschatological promises of Christianity would be fulfilled or even eclipsed. Often resorting to the vocabulary of traditional religion and myth or to images taken from them, the ideologues of extreme progress emphasise that humanity or its (biological or artificial) offspring will legitimately assume a God-like role. The visions of human enhancement serve only to define and outline the first steps that are to be taken in order to reach this glorious distant future: only radically modified human beings ('transhumans') or beings that are wholly artificial ('posthumans') will be able to conquer and colonise space.

With very few exceptions, the ideologues refrain from any vision of a final culmination of progress, retaining instead a modern notion of infinite progress. They systematically weaken the authentic rationality of the modern idea of progress, however, particularly when the far-reaching visions of human enhancement and posthumanity take centre stage in their discourse. The ideology of extreme progress thus offers a specific answer to a structural problem of modernity, namely the overstraining of progressive rationality by eschatological hopes, and is at the same time a symptom of this overstraining. In a Kantian vein, Blumenberg wrote the following about progress:

> Infinite progress does make each present relative to its future, but at the same time it renders every absolute claim untenable. This idea of

progress corresponds more than anything else to the only regulative principle that can make history humanly bearable, which is that all dealings must be so constituted that through them people do not become mere means. (Blumenberg, 1966, p. 35)

In their grand schemes of the future, Wells, Haldane and Bernal did, on various occasions, imagine individuals as mere means for the further progress of humanity, as in the case of Bernal's bizarre utopian zoo or Haldane's notions that '[t]he average man...must learn that the highest of his duties is to assist those who are creating, and the worst of his sins to hinder them' and that 'the value of the individual is negligible in comparison with' the 'cosmic destiny of mankind' (Haldane, 1927a, p. 39f.). Even when they emphasised again and again, in particular from the 1930s onwards, that science should be placed in the service of society, one often gains the impression that they wanted to see this goal realised only in order to facilitate thereafter the total submission of society to science. This total submission could appear to be necessary in order to make possible the progress of science which Haldane had defined in his first futuristic essay as 'man's gradual conquest, first of space and time, then of matter as such, then of his own body and those of other living beings, and finally the subjugation of the dark and evil elements in his own soul' (Haldane, 1923, p. 81f.) – Bernal's 'world', 'flesh' and 'devil' (Bernal, 1929). Since the human race has to 'prove that its destiny is in eternity and infinity' (Haldane, 1927a, p. 40), it has to conquer outer space to avoid ultimate destruction. Nevertheless, their elitist and technocratic – and at the same time eschatological – ideas were always counterbalanced at least by a genuine interest in rational and socially useful applications of science in the here and now. To most of their friends, admirers and opponents, at least from the 1930s onwards, their far-reaching visions only appeared as a curious or fascinating accessory of their social and political radicalism.

In the absence of strong faith in social progress, however, and for as long as the daily vilification of humanity and reason all over the world continually derides even such an unassuming definition of progress as that given by Blumenberg, radical visions of human enhancement will definitely not be a corollary phenomenon of social progressivism. On the contrary, they serve anti-progressive interests, being a symptom of the overstraining of progressive rationality rather than its culmination. As elements of an explicitly techno-eschatological worldview, these visions are in fact an opiate – if not yet for the masses, then surely for important functional elites within the middle classes, including a considerable number of scientists and engineers. There is reason to fear that this will continue to severely hamper rational discourse on emerging bio- and neurotechnologies in the future (for the following, see also Coenen, 2009). Since these visions appear to be largely unrealistic, they serve as propagandistic goals which camouflage the much more mundane interests of the involved parts of the science system. This is facilitated by the fact that the visions that were created or popularised by the ideologues of extreme progress have exerted a significant influence on the sociotechnical imagination in our societies, above all through the dystopian reactions to them and *via* the genre of science fiction.

Because of its origins in the science system, the ideology of extreme progress can be deemed an enemy of science from within. The highly visionary discourses on human enhancement and other discourses influenced by this ideology suffer from an ideological imbalance: at one end of the spectrum of opinions, the one marked by radical criticisms of technology and progress, a barrier is erected against unscientific beliefs and fundamentalist currents of thought. At the other end of the spectrum, however, the one characterised by far-reaching visions of technoscientific progress, the boundaries separating these visions from salvation ideologies and mythical thought are permeable. Visionary marketing contributes to the decomposition of scientific rationality under a technoscientific regime prone to irrational communication and dealing in supposed technological fixes for eschatological fears.

References

Anders G. (1956) *Die Antiquiertheit des Menschen. Über die Seele im Zeitalter der zweiten industriellen Revolution* (München: Beck, 2002).

Bernal J.D. (1929) *The World, the Flesh and the Devil. An Inquiry into the Future of the Three Enemies of the Rational Soul* (London: Jonathan Cape, 1970).

Blumenberg H. (1966) *The Legitimacy of the Modern Age* (Cambridge MA and London: MIT Press, 1983).

Butler S. (1872) *Erewhon: or Over the Range* (London: Trubner and Co.).

Coenen C. (2009) 'Zauberwort Konvergenz', *Technikfolgenabschätzung – Theorie und Praxis (TATuP)*, 18(2), 44–50.

Coenen C. (2010) 'Immagini di società potenziate dalla nanotecnologia. L'ascesa dell'ideologia postumanista del progresso estremo', Amistani F. and Arnaldi S. (transl.), in Arnaldi S. and Lorenzet A. (eds) *Innovazione in corso. Il dibattito sulle nanotecnologie fra diritto, etica e società* (Bologna: Il Mulino), pp. 225–58.

Coenen C., Schuijff M., Smits M., Pim K., Hennen L., Rader M., Wolbring G. (2009) *Human Enhancement* (Brussels: European Parliament), IP/A/STOA/FWC/2005-28/SC32 & 39, www.europarl.europa.eu/stoa/publications/studies/stoa2007-13_En.pdf.

Darwin C. (1871) *The Descent of Man, and Selection in Relation to Sex* (London: John Murray), Vol. 2, first edn, scanned by John van Wyhe, http://darwin-online.

org.uk/content/frameset?itemID=F937.2&viewtype=text&pageseq=1 Accessed September 9, 2013.
Gauchet M. (1985) *The Disenchantment of the World. A Political History of Religion* (Princeton: Princeton University Press, 1997).
Haldane J.B.S. (1923) *Daedalus or Science and the Future* (a paper read to the heretics, Cambridge, 4 February 1923) (London: Kegan Paul, Trench, Trubner & Co., 1925).
Haldane J.B.S. (1927a) *The Last Judgment. A Scientist's Vision of the Future of Man* (New York and London: Harper and Brothers).
Haldane J.B.S. (1927b) 'Man's destiny', in Haldane J.B.S., *The Inequality of Man* (Harmondsworth: Pelican Books, 1932, 2nd edn 1937), pp. 140–5.
Haldane J.B.S. (1932) 'Possibilities of human evolution', in Haldane J.B.S., *The Inequality of Man* (Harmondsworth: Pelican Books, 1932, 2nd edn 1937), pp. 82–99.
Hillegas M. (1967) *The Future as Nightmare. H. G. Wells and the Anti-Utopians* (New York: Oxford University Press).
Hogan D. and Clarfield M. (2007) 'Venerable or vulnerable: ageing and old age in JRR Tolkien's *The Lord of the Rings*', *Medical Humanities* 33, 5–10.
Huxley A. (1932) *Brave New World* (London: Chatto & Windus).
Khushf G. (2005) 'The use of emergent technologies for enhancing human performance: are we prepared to address the ethical and political issues?', *Public Policy & Practice* (e-journal), 4(2), n.p., http://www.ipspr.sc.edu/ejournal/Archives0805.asp.
Kurzweil R. (2005) *The Singularity Is Near* (New York: Viking Penguin).
Löwith K. (1949) *Meaning in History* (Chicago and London: The University of Chicago Press)
Meadows J. (2004) *The Victorian Scientist. The Growth of a Profession* (London: The British Library).
Nordmann A. (2008a) 'Ignorance at the heart of science', in Ach J.S. and Lüttenberg B. (eds) *Nanobiotechnology, Nanomedicine and Human Enhancement* (Berlin: LIT Verlag), pp. 113–32.
Nordmann A. (2008b) 'No future for nanotechnology? Historical development vs global expansion' in Jotterand F. (ed.) *Emerging Conceptual, Ethical and Policy Issues in Nanobiotechnology* (Netherlands: Springer), pp. 43–63.
Parrinder P. (1995) *Shadows of the Future. H.G. Wells, Science Fiction and Prophecy* (Liverpool: Liverpool University Press).
Pope A. (1733) *An Essay on Man. Addressed to a Friend*. Part 1 (London: J. Wilford).
Reade W.W. (1872) *The Martyrdom of Man* (Chestnut Hill: Adamant Media Corporation, 2005).
Saunders M. and Hurwitz B. (2009) 'The To-day and To-morrow Series and the popularization of science: an introduction', *Interdisciplinary Science Reviews*, 34(1), 3–8.
Schneider S. (2009) 'Future minds: transhumanism, cognitive enhancement, and the nature of persons', in Ravitsky V., Fiester A. and Caplan A.L. (eds) *The Penn Center Guide to Bioethics* (New York: Springer), pp. 95–110.
Slusser G. (2009) 'Dimorphs and doubles: J. D. Bernal's "Two cultures" and the transhuman promise', in Westfahl G. and Slusser G. (eds) *Science Fiction and the*

Two Cultures. Essays on Bridging the Gap between the Sciences and the Humanities (Jefferson: McFarland & Co.), pp. 96–130.

Stableford B. (1985) *Scientific Romance in Britain 1890–1950* (London: Fourth Estate).

Taylor C. (2007) *A Secular Age* (Cambridge MA: Harvard University Press).

Wells H.G. (1902) *The Discovery of the Future* (A discourse delivered to the Royal Institution on 24 January 1902) (London: T. Fisher Unwin).

10
Transcendental Medicine versus the 'Prisonhouse of the Flesh': Enhancement in R.L. Stevenson's *The Strange Case of Dr Jekyll and Mr Hyde*

Françoise Dupeyron-Lafay

'L'homme n'est ni ange ni bête, et le malheur veut que qui veut faire l'ange fait la bête'

(Pascal, 1670, Fragment 572, p. 370).[1]

Robert Louis Stevenson's *The Strange Case of Dr Jekyll and Mr Hyde* (Stevenson, 1886) is a world famous fantasy text presenting a respectable Victorian doctor's personality splitting in two thanks to a draught that results in the embodiment of his other evil self. Partly because of film adaptations that schematise and stylise the contents to make it more sensational and visual, the spectacular transformation of the doctor in his laboratory and the havoc wreaked by Hyde are what dominates in the public's mind. However, Stevenson's text is more subtly uncanny and provides an in-depth and complex theorisation of the psychological and metaphysical significance and origins of the double.

As a matter of fact, *The Strange Case* only briefly touches upon the scientific side itself whereas the laboratory apparatus is often featured prominently and sensationally in film versions. The narrative remains rather vague about the draught or potion and its components, for one thing because, unlike H. G. Wells, Stevenson had had no scientific training, but above all, because this was not his priority. The crux of the matter was actually the question of a yearned-for (personal and collective) 'human enhancement', a very modern term that may initially seem quite anachronistic for a late-Victorian text – all the more as it is a work of fiction and not a scientific essay. But the term stems from an age-old human longing that describes Jekyll's project very aptly, namely his desire to break free from the shackles of his imperfect humanity and improve morally, and to a lesser extent physically, by creating a double that would ideally embody and bear the blame for the worse aspects of his self. Contemporary medicine was of course radically improving (technically) but there was a growing feeling of dissatisfaction with materialism and the mechanistic and naturalistic paradigms of the world and of man offered by science. This accounts for the prominent role of 'transcendental medicine' as the only (supposedly) viable way to human enhancement. This enigmatic, never clearly defined notion seems to be a contradiction in terms; however, far from being an oxymoron, it postulates reconciliation and unity, and in Stevenson's novella, it possesses an obviously holistic power and aura.

Human enhancement in *The Strange Case* involves dramatic physical and psychological transformations induced by the use of psychotropic substances, the origins of which are presumably autobiographical. In the 1880s, Stevenson, whose health had been precarious from childhood, took laudanum, also known as 'tincture of opium', a psychotropic drug that belongs to the category of 'narcotics', particularly to relieve the pain caused by lung haemorrhages and to fight against sleeplessness. Laudanum could have served as an inspiration for Jekyll's draught and may account for the dark dreamlike quality of *The Strange Case*. However, according to Pascal Nouvel (2010), cocaine, which Stevenson also used, may have represented another model for the doctor's psychotropic substance – although he denied ever resorting to any substance resembling Jekyll's potion when interviewed by a journalist. Nouvel emphasises the sedative properties of laudanum as opposed to the stimulating effects of cocaine, expounded in Myron Schultz's paper 'The Strange Case of Robert Louis Stevenson: a Tale of Toxicology', published in the *Journal of the American Medical Association* in 1971. Indeed, Schultz argues that cocaine is a very likely model for Dr Jekyll's potion, as Stevenson wrote the story so quickly, despite his physical weakness, and declared being exhilarated, instead of exhausted, by the experience. Cocaine had been made available for medical use in the 1880s and many doctors, among them Sigmund Freud, were in favour of it. Nouvel mentions Freud's account of his personal use of it in *Über Coca*, that shows its intensely stimulating and exhilarating effect, and can be paralleled with Jekyll's description of his sensations when he takes the drug for the first time. He points out the 'novelty' and 'the freshness' of what he experienced and the fact he suddenly 'felt younger, lighter, happier in body' (Stevenson, 1886, p. 62). Whatever the medical data on cocaine or its possible influence in Stevenson's novella, my point is that laudanum (and opium) can be described as having exactly similar effects.[2] My contention, therefore, is that Stevenson did not use a single model for the description of the physiological and psychological effects of Jekyll's potion, but a compound one, inspired from his own 'human-enhancing' experiences with both laudanum and cocaine.

If the role of psychotropic substances should not be underestimated, it is the strongly Calvinistic context which illuminates their meaning in the novella that hinges primarily on ethical questions. It is divided into ten parts: the last two are retrospective and consist of two letters read posthumously, respectively, 'Doctor Lanyon's Narrative' that explicitly presents Jekyll's metamorphosis to the reader for the first time, and 'Doctor Jekyll's Full Statement of the Case', that minutely explains the meaning and purpose of the transformation, namely the moral improvement of man, and the attempt to isolate his inherently and inevitably bad side from his better self. However, in the novella, this cannot be achieved without the radical physical separation of these good and evil sides coexisting uneasily in each individual. The story goes on to explore how to deal with and control the undesirable double, the sheer and tragic impossibility of which leads to Jekyll's moral and physical downfall and destruction.

I should therefore like to examine the origins and aims of enhancement in the text, then its modalities and finally its disastrous results and the causes for this failure that borrows many traits from classical tragedy.

The origins and aims of enhancement within a Calvinistic context

A few preliminary biographical facts concerning Stevenson can shed light on some key aspects of the text, although *The Strange Case* is not *directly* autobiographical. The novella is supposed to take place in London (Hyde lives in the district of Soho, for instance) but many critics agree that the dark, gloomy and foggy setting of the story rather evokes Edinburgh where Stevenson was born in 1850 and spent his youth. Both his parents were fervent Calvinists, and from his childhood, he suffered from lung problems and weak health because the Scottish weather did not agree with him. He had to be kept indoors for whole weeks in winter, hence his escape into fantasy. His imagination was nourished by his nurse, Alice Cunningham, who told him many tales. As he could not attend school regularly, he was also an avid reader. He must have been familiar from childhood with the story of Deacon Brodie, a respectable Scottish cabinet maker during the daytime and a burglar at night. Stevenson was supposed to become an engineer like his father but became a law student instead, and although he was not interested in the legal field, he managed to pass his exam and to be admitted to the Scottish Bar. While still at University, Stevenson started going to shady places, leading a Bohemian life, associating with disreputable people and having an affair with a prostitute. His parents, of course, opposed a wedding. His literary career, which started in the mid 1870s, was fostered by these various experiences, but it may not have been as rich and productive without the drugs Stevenson took from the 1880s onwards.

Besides, the burden of religion and the role of the law in Stevenson's life account for the importance of the character of Utterson in the novella. This very taciturn and austere – albeit friendly and tolerant – bachelor is a lawyer, and he plays a key role in the story, both as the main witness ensuring the coherence of the whole and as an amateur detective.

Indeed, the world of *The Strange Case* is exclusively masculine: all the main characters (Dr Jekyll, Dr Lanyon, Utterson and his cousin Mr Enfield) are bachelors and solitary men, and the portrait of Utterson in Chapter 1 epitomises religious strictures and Victorian repression, also foreshadowing the more radical situation of his friend and client, Dr Jekyll:

> He was austere with himself; drank gin when he was alone, to mortify a taste for vintages; and though he enjoyed the theatre, had not crossed the doors of one for twenty years. But he had an approved tolerance for others; sometimes wondering, almost with envy, at the high pressure of spirits involved in their misdeeds; and in any extremity inclined to help rather than to reprove. 'I incline to Cain's heresy,' he used to say quaintly: 'I let my brother go to the devil in his own way'. In this character, it was frequently his fortune to be the last reputable acquaintance and the last good influence in the lives of down-going men. (Stevenson, 1886, p. 7)

The introduction to Utterson clearly evokes the puritanical Scottish context in which respectability and pleasure, even in its comparatively innocent forms, were regarded as incompatible. This divorce is at the heart of Dr Jekyll's tragedy and represents the origin of the split between his true face or self and his social mask; let us bear in mind he is called 'the very pink of proprieties' (Stevenson, 1886, p. 11). In his final confession, or 'Full Statement of the Case', Jekyll himself shows that he had long suffered from this policy of concealment that paves the way for the existence of 'Mr Hyde':

> And indeed the worst of my faults was a certain impatient gaiety of disposition, such as has made the happiness of many, but such as I found it hard to reconcile with *my imperious desire to carry my head high, and wear a more than commonly grave countenance before the public.* Hence it came about that I concealed my pleasures; and that when I reached years of reflection, and began to look round me and take stock of my progress and position in the world, I stood already committed to a profound duplicity of life. Many a man would have even blazoned such irregularities as I was guilty of; but from *the high views that I had set before me*, I regarded and *hid them with an almost morbid sense of shame.* (Stevenson, 1886, p. 60; emphasis added)

The theme of duality and the double was not new but Stevenson's work gives it a new form in *The Strange Case*, especially by showing the baneful effects of repression among would-be representatives of normality and order such as lawyers or doctors.[3] In Poe's 'William Wilson' (Poe, 1839) or Stevenson's previous story entitled 'Markheim' (Stevenson, 1884), the doppelgänger had been the embodiment of the protagonist's conscience or better self. Moreover, the two William Wilsons had been exact replicas while Hyde looks radically different, bears his own name, has his own handwriting and signature, and a house of his own. Moreover, in his 'statement', Jekyll even comes to doubt the fact man is merely double and his conclusion heralds psychoanalytical research:

> I thus drew steadily nearer to that truth, by whose partial discovery I have been doomed to such a dreadful shipwreck: that man is not truly one, but truly two. I say two, because the state of my own knowledge does not pass beyond that point. Others will follow, others will outstrip me on the same lines; and I hazard the guess that man will be ultimately known for a mere polity of multifarious, incongruous, and independent denizens. (Stevenson, 1886, p. 61)

The Strange Case is an unmistakable Victorian product and a cultural phenomenon in which religious and ethical questions coexist uneasily with science. In his 'statement', Jekyll obviously evokes his personal case, insisting that he was no worse than his fellow men but that he had a more acute conscience and a stronger superego:

> It was thus rather the exacting nature of my aspirations than any particular degradation in my faults, that made me what I was and, with even a deeper trench than in the majority of men, severed in me those provinces of good and ill which divide and compound man's dual nature. (Stevenson, 1886, p. 60)

The root of his predicament did not lie solely in himself: it was universal and grounded in the scriptural conception of man as essentially corrupt and sinful. This form of determinism and the insufferable burden of his religious upbringing trigger the doctor's tragedy and make him unable to bear this 'thorough and primitive duality of man' (Stevenson, 1886, p. 61). For Jekyll, religion is nothing but 'one of the most plentiful springs of distress' because its very principle is 'that hard law of life' (Stevenson, 1886, p. 60), namely 'the curse of mankind' (Stevenson, 1886, p. 61) whereby every man is bound to be partly evil and partly good, something Utterson seems to accept when he evokes his tolerance of 'Cain's heresy'. But Jekyll cannot come to terms with it, and the only way out for him is science and 'transcendental medicine' as the 'direction of [his] scientific studies...led wholly toward the mystic and the transcendental' (Stevenson, 1886, p. 60). Therefore, the doctor initially sees his research as a remedy with a miraculous cathartic potential able to free the self from its moral hybridity and thereby to exorcise evil:

> even before the course of my scientific discoveries had begun to suggest the most naked possibility of such a miracle, I had learned to dwell with pleasure, as a beloved day-dream, on the thought of the separation of these elements. (Stevenson, 1886, p. 61)

The very sophisticated writing style, especially all the symbolic and metaphoric devices present in Jekyll's 'statement', involves the association of duality with vertical dynamics: the 'more upright twin might go his way...on his upward path' while the 'extraneous evil' (Stevenson, 1886, p. 61) embodied by Hyde is later referred to as 'the slime of the pit' (Stevenson, 1886, p. 74).

We can trace back the origin of the rich metaphoric network in the text to the Biblical background,[4] and to the specific nature of the medicine favoured by Jekyll. Because of his thorough knowledge of Scripture, he views the body as something merely transient and as flimsy as a garment. As we can read in the *King James Bible*: 'Unto Adam also and to his wife did the LORD God make coats of skins, and clothed them' (Genesis 3:21); and 'Is not the life more than meat, and the body than raiment?' (Matthew 6:25). The metaphoric network that equates the self and a piece of clothing is a direct scriptural echo, when Jekyll writes:

> I was so far in my reflections when, as I have said, a side-light began to shine upon the subject from the laboratory table. I began to perceive more deeply than it has ever yet been stated, the trembling immateriality, the mist-like transience of this seemingly so solid body in which we walk attired. Certain agents I found to have the power to shake and to pluck back that fleshly vestment, even as a wind might toss the curtains of a pavilion. (Stevenson, 1886, p. 61)

Accordingly, the adventurous doctor can put on and take off his respectable or his dark self at will, or so he thinks:

> I had but to drink the cup, to doff at once the body of the noted professor, and to assume, like a thick cloak, that of Edward Hyde. (Stevenson, 1886, p. 64)

The way the body and identity are referred to is quite fluctuating and ambiguous, probably reflecting the inner split within the subject: identity is defined as a 'fortress' (Stevenson, 1886, p. 62), a term expressing the notions of safety and a theoretically impregnable refuge, and almost in the same breath it is said to be 'an immaterial tabernacle' (Stevenson, 1886, p. 62), hence a holy and sacred place (through the reference to the tent holding the Ark of the Covenant and all the sacred vessels before the building of the Temple), but also something quite insubstantial. However, Jekyll also sees his disposition as a 'prisonhouse' (Stevenson, 1886, p. 64). The *Book of Genesis* serves as a strong inspiration for some of the other powerful metaphors used by Jekyll in his 'statement'. His two selves are described as 'polar twins' (Stevenson, 1886, p. 61) and enemy brothers. The conflict is in fact an internecine strife represented by 'the two natures that contended in the field of [his] consciousness' (Stevenson, 1886, p. 61). Jekyll goes on to add:

> It was the curse of mankind that these incongruous fagots were thus bound together – that in the agonised womb of consciousness, these

polar twins should be continuously struggling. How, then, were they dissociated? (Stevenson, 1886, p. 61)

Once more, the answer to that question lies in the Bible:

> And if thy right hand offend thee, cut it off, and cast it from thee: for it is profitable for thee that one of thy members[5] should perish, and not that thy whole body should be cast into hell. (Matthew, 5:30)

Nevertheless, Hyde is not the *right* hand but the 'sinister', or *left* one. The comparison between Jekyll's and Hyde's handwritings is telling. As Mr Guest (Utterson's head clerk) remarks: 'the two hands are in many points identical: only differently sloped' (Stevenson, 1886, p. 34). Jekyll gives us the key, explaining that 'by sloping [his] own hand backwards, he had supplied [his] double with a signature' (Stevenson, 1886, p. 66). Sloping backwards obviously (and symbolically) means 'to the left'.

Externalising the double serves to purify the better self and purge him from the unacceptable evil within himself, as Otto Rank makes it clear in *The Double* (Otto Rank, 1914, initially published in *Imago*), an essay that encompasses both the fields of anthropology and literature. In Chapter 4, devoted to the 'The Shadow as the representation of the soul', Rank shows how the status of the double shifted from protective entity or guardian angel ensuring immortality, to persecutor and tormentor. Chapter 5, 'The Reflection as a symbol of narcissism', evokes the defence mechanisms whereby the individual divides itself from that part of its self it cannot come to terms with. And Chapter 7, 'The Belief in the immortality of the self', very aptly describes Jekyll's situation and his blaming his evil and sinful acts or tendencies on another. Rank explains that these negative elements are severed from the self and integrated to the double, thereby enabling the 'hero' to indulge them without the necessity to shoulder the responsibility. Indeed, Jekyll refers to his evil side as 'this extraneous evil' (Stevenson, 1886, p. 61), but when he first transforms himself, he initially recognises and accepts his projected self:

> And yet when I looked upon that ugly idol in the glass, I was conscious of no repugnance, rather of a leap of welcome. This, too, was myself. It seemed natural and human. In my eyes it bore a livelier image of the spirit, it seemed more express and single, than the imperfect and divided countenance I had been hitherto accustomed to call mine. (Stevenson, 1886, p. 63)

However, he soon rejects it, calling it 'my vicarious depravity' (Stevenson, 1886, p. 65). And he exculpates himself from the atrocities committed by his other self:

> It was Hyde after all, and Hyde alone, that was guilty. Jekyll was no worse; he woke again to his good qualities seemingly unimpaired; he would even make haste, where it was possible, to undo the evil done by Hyde. (Stevenson, 1886, p. 66)

It is quite striking to see he also starts speaking of himself in the third person as Henry Jekyll and/or Edward Hyde. Later on, he insists, he cannot call his other self 'I' (Stevenson, 1886, p. 73), but he is no longer able to refer to his better self as 'I' either, and his detached consciousness hovers uneasily between them. Symbolically, before the metamorphoses, he had no looking-glass in his room and did not know himself.[6] Indeed, when he first becomes Hyde, he regards himself as 'a stranger in [his] own house' (Stevenson, 1886, p. 63) that no longer belongs to him (Stevenson, 1886, p. 64)[7]. We should bear in mind Hyde owns his own house in Soho just as Jekyll's experiment has succeeded in 'hous[ing]' each of the two principles (good and evil) 'in separate identities' (Stevenson, 1886, p. 61). Nevertheless, Hyde owns a key to Jekyll's house and gets free access to it. This gains added meaning in the light of Freud's concept of the mind as a divided house inhabited by distinct agencies in 'General Theory of the Neuroses', the third part of *A General Introduction to Psychoanalysis* (Freud, 1916–17, Chapter 19, 'Resistance and suppression'). One of Freud's key notions and his justification for the existence of psychoanalysis is that the self is not even master of/in its own house (Chapter 18, 'Traumatic fixation – the Unconscious'). *The Strange Case* provides an early literary illustration of this by showing that the house and the self, as mirror-images of each other, were not the inviolable sanctuaries they had long been thought to be, and that they sheltered alien, disruptive and conflicting agencies.

Pascal Nouvel (2010) traces the influence of scientific literature upon the 'core idea' of the novella, quoting Fanny Osbourne-Stevenson's introduction to her husband's collected works in which she showed how 'deeply impressed' he had been by a paper on the subconscious he had read in a French scientific journal, and how this held the seeds for Jekyll and Hyde. Nouvel evokes Stevenson critics such as Jacqueline Carroy, who believes the paper evoked by Fanny was in fact a series of texts written by Eugène Azam in *La Revue scientifique* in the 1870s, focusing on Azam's patient Felida X., whom Ian Hacking identifies as 'the first French double personality to be studied in detail'.

The case studies of Felida were translated and summarized by Richard Proctor in a series of articles that were published in *The Cornhill Magazine* in the late 1870s. Since Stevenson also contributed to *The Cornhill Magazine* at the time, it is very unlikely that these articles escaped his attention. Thus, the cases of double personality as they were extensively discussed are very good candidates for having been sources for Stevenson's Jekyll and Hyde character. (Nouvel, 2010)

The modalities of the transformations: a monstrous scenario of birth-giving. Regression and degeneration

The way Dr Jekyll achieved his transformation and the exact nature of his scientific procedure remain nebulous and shrouded in mystery, which heightens the sense of uncertainty and terror. Lanyon's description of the ingredients of the draught does not clarify anything but sounds rather ominous with its mention of 'blood-red':

> The phial, to which I next turned my attention, *might have been about* half-full of a blood-red liquor, which was highly pungent to the sense of smell and *seemed to me to* contain phosphorus and *some* volatile ether. At the other ingredients *I could make no guess.* (Stevenson, 1886, p. 55; emphasis added)

The only clue given us is that the transforming draught has psychotropic properties that are so powerful that they also modify the body. But as I pointed out in the introduction, this is not what really matters in the novella: 'For two good reasons, I will not enter deeply into this scientific branch of my confession' (Stevenson, 1886, p. 61). Jekyll contents himself with making excuses for this absence of definite information and remarks:

> [I] managed to compound a drug by which these powers should be dethroned from their supremacy, and a second form and countenance substituted, none the less natural to me because they were the expression, and bore the stamp, of lower elements in my soul.
>
> I hesitated long before I put this theory to the test of practice. I knew well that I risked death; for any drug that so potently controlled and

shook the very fortress of identity, might by the least scruple of an overdose or at the least inopportunity in the moment of exhibition, utterly blot out that immaterial tabernacle which I looked to it to change. But the temptation of a discovery so singular and profound, at last overcame the suggestions of alarm. I had long since prepared my tincture; I purchased at once, from a firm of wholesale chemists, a large quantity of a particular salt which I knew, from my experiments, to be the last ingredient required; and late one accursed night, I compounded the elements, watched them boil and smoke together in the glass, and when the ebullition had subsided, with a strong glow of courage, drank off the potion. (Stevenson, 1886, p. 62)

The doctor's vagueness or reticence may then have obvious autobiographical origins as Stevenson was a regular – if not addicted – drug user, but would not own up to it. Besides, the writer wished to preserve the complexity and ambiguity of his story, and to avoid clear-cut, simplistic interpretations of it, so that it should not be read as a mere scientific tale.[8] In the same way, the quarrel between the two medical schools or philosophies is only very briefly evoked. The two conflicting approaches are embodied by the two doctors of the story, namely Jekyll and Dr Lanyon whom Jekyll accuses of having 'so long been bound to the most narrow and material views' and of having 'denied the virtue of transcendental medicine' and 'derided [his] superiors' (Stevenson, 1886, p. 58). The exact meaning of 'transcendental medicine' is never fully specified, mainly because of its miraculously holistic properties and its unprecedented, wish-fulfilling, 'human-enhancing' power; its delineation is actually achieved in contradistinction to the purely physiological, mechanistic, materialistic type of medicine that Lanyon advocates, divorcing body, mind and 'soul'. Ironically, Jekyll's holistic medical approach, which should promote unity and overall progress, rests on an initial division of the two selves, and leads to an irrevocable and fatal split.

It is through Lanyon's posthumous letter that the reader is belatedly and for the first time informed of what Jekyll was doing. Indeed, 'Doctor Lanyon's narrative' stages Hyde's visit to him and his staggering metamorphosis into Jekyll, a shock that will finally kill Lanyon over the following fortnight. This is probably the most spectacular moment in the book, as the scene is happening in real time in front of Lanyon and it is dramatised so efficiently that it seems to be taking place in front of us, too:

He put the glass to his lips and drank at one gulp. A cry followed; he reeled, staggered, clutched at the table and held on, staring with

> injected eyes, gasping with open mouth; and as I looked there came, I thought, a change – he seemed to swell – his face became suddenly black and the features seemed to melt and alter – and the next moment, I had sprung to my feet and leaped back against the wall, my arm raised to shield me from that prodigy, my mind submerged in terror.
>
> 'O God!' I screamed, and 'O God!' again and again; for there before my eyes – pale and shaken, and half-fainting, and groping before him with his hands, like a man restored from death – there stood Henry Jekyll! (Stevenson, 1886, pp. 58–9)

Just as the relationship between good and evil is unstable and ambiguous, life coexists with death through each transformation. The metamorphosis passes through a stage reminiscent of death throes, and it involves racking pains before the other being can be born or reborn. In the final section of the novella, following Dr Lanyon's narrative, Dr Jekyll presents the same process but in reverse, with the first description of his transformation into Hyde and a stronger emphasis on the notions of (re)birth and liberation:

> The most racking pangs succeeded: a grinding in the bones, deadly nausea, and a horror of the spirit that cannot be exceeded at the hour of birth or death. Then these agonies began swiftly to subside, and I came to myself as if out of a great sickness. There was something strange in my sensations, something indescribably new and, from its very novelty, incredibly sweet. I felt younger, lighter, happier in body; within I was conscious of a heady recklessness, a current of disordered sensual images running like a mill-race in my fancy, a solution of the bonds of obligation. (Stevenson, 1886, p. 62)

Each new appearance of Hyde seems to be the result of a preternatural scenario of childbirth or delivery which had been symbolically prepared by the metaphor of the 'polar twins' struggling in the 'agonised womb of consciousness' (Stevenson, 1886, p. 61). By the end of his confession, the doctor represents the antagonistic relationship of Jekyll and Hyde in a way that evokes the simultaneous creation of Adam and Eve as one body and one flesh with Eve–Hyde violently struggling to get free:

> This was the shocking thing; that the slime of the pit seemed to utter cries and voices; that the amorphous dust gesticulated and sinned; that what was dead, and had no shape, should usurp the offices of

life. And this again, that that insurgent horror was knit to him *closer than a wife*, closer than an eye; lay caged in his flesh, where he heard it mutter and felt it struggle to be born. (Stevenson, 1886, p. 75; emphasis added)

Further, Hyde is described as smaller and younger, as if he were Jekyll's child and, initially, the doctor describes his early metamorphoses in an almost playful tone, emphasising the notion of a boyish prank:

I was the first that could thus plod in the public eye with a load of genial respectability, and in a moment, like a schoolboy, strip off these lendings and spring headlong into the sea of liberty. But for me, in my impenetrable mantle, the safety was complete. (Stevenson, 1886, p. 65)

But the transformation means much more than this as 'all human beings, as we meet them, are commingled out of good and evil: and Edward Hyde, alone in the ranks of mankind, was pure evil' (Stevenson, 1886, p. 64). Jekyll has given birth to a son that does not care for him but merely needs him:

Hyde was indifferent to Jekyll, or but remembered him as the mountain bandit remembers the cavern in which he conceals himself from pursuit. Jekyll had more than a father's interest; Hyde had more than a son's indifference. (Stevenson, 1886, p. 68)

It is only because 'The evil side of [Jekyll's] nature...was less robust and less developed than the good which [he] had just deposed' (Stevenson, 1886, p. 63) that Hyde remains provisionally smaller but he soon grows in strength and stature, and Jekyll evokes his own spontaneous and uncontrolled transformations into his evil alter ego without any need of the drug. Hyde finally tries to get the upper hand, to supplant and kill his 'father' Jekyll, and even his 'grandfather':

And certainly the hate that now divided them was equal on each side. With Jekyll, it was a thing of vital instinct. He had now seen the full deformity of that creature that shared with him some of the phenomena of consciousness, and was co-heir with him to death: and beyond these links of community, which in themselves made the most poignant part of his distress, he thought of Hyde, for all his energy of life, as of something not only hellish but inorganic.... The hatred of Hyde for Jekyll, was of a different order. His terror of the gallows drove him continually to commit temporary suicide, and return to his subordinate station of a part instead of a person; but he loathed the necessity, he loathed the despondency into which Jekyll was now fallen, and he resented the dislike with which he was himself regarded. Hence the ape-like tricks that he would play me, scrawling in my own hand blasphemies on the pages of my books, burning the letters and destroying the portrait of my father; and indeed, had it not been for his fear of death, he would long ago have ruined himself in order to involve me in the ruin. (Stevenson, 1886, pp. 74–5)

This independent double persecutes his 'father'. The hate and loathing of Jekyll for Hyde and the creature's resentment – 'he loathed the despondency into which Jekyll was now fallen, and he resented the dislike with which he was himself regarded' (Stevenson, 1886, p. 75) – are reminiscent of Victor Frankenstein's relationships with the 'monster'[9] he created. But paradoxically, though Hyde looks younger, he is also much older: he is called 'troglodytic' (Stevenson, 1886, p. 19) in the second section of the novella ('Search for Mr Hyde') because of his primitive state as a kind of caveman, and also because Jekyll, as we saw, is a 'cavern' (Stevenson, 1886, p. 68) for him. The very aptly named Hyde embodies a terrifying return of the repressed and of primitive drives, a form of regressive otherness, as the recurrent devolutionist references to his simian appearance and behaviour testify. The ape has probably much to do with Darwin's theory of 'descent with modification'. During the brutal night murder of the MP in the fourth section ('The Carew Murder Case'), he tramples his victim and clubs him to death 'with ape-like fury' (Stevenson, 1886, p. 26). Jekyll twice uses the same adjective in his own narrative when he remembers the 'apelike tricks' Hyde played him (Stevenson, 1886, p. 75), and the 'apelike spite' he kept showing (Stevenson, 1886, p. 76).[10] Hyde's bestiality and violence are recurrently mentioned by the doctor, especially before the murder of Carew: 'My devil had been long caged, he came out roaring' (Stevenson, 1886, p. 69). Each new sinful act leads Jekyll to recoil in horror and feel deep contrition, but he cannot resist the call of the animal within:

> ...as the first edge of my penitence wore off, the lower side of me, so long indulged, so recently chained down, began to growl for licence. (Stevenson, 1886, p. 71)

Any contact with Hyde, and even the very sight of him, are morally harmful: he seems to corrupt those he meets as if his evil rubbed off onto them or, worse still, aroused and unleashed each one's worse instincts. The first section of the book, 'Incident of the Door' staging Hyde's cruelty and violence for the first time (he tramples a little girl) is a case in point. The scene turns the people present into potential murderers, including the apparently impassive doctor called to attend the child, and Mr Enfield, the narrator and main witness of the scene. Like his cousin Mr Utterson, this man is normally a paragon of virtue and uprightness. In the end, the furious crowd that has captured Hyde has to be restrained:

> I saw that Sawbones turn sick and white with the desire to kill him. I knew what was in his mind, just as he knew what was in mine; and killing being out of the question, we did the next best.... And all the time... we were keeping the women off him as best we could, for they were as wild as harpies. I never saw a circle of such hateful faces. (Stevenson, 1886, p. 10)

The city itself is depicted in uncanny, animal terms with its 'low growl' 'all around' (Stevenson, 1886, p. 17), suggesting the existence of pent-up disruptive forces that could easily break loose.

Ironically and tragically, human enhancement leads to physical and moral degradation in Stevenson's story as the very first transformation, although welcomed with joy, clearly suggests:

> I stretched out my hands, exulting in the freshness of these sensations; and in the act, I was suddenly aware that I had *lost in stature*. (Stevenson, 1886, p. 63; emphasis added)

The disastrous results of 'enhancement' and the causes for this tragic failure. Universal imperfection and impurity as the tragic flaws of the story

Dr Jekyll is not the only 19th-century scientist in literature. He is preceded by Victor Frankenstein (Mary Shelley, 1818), or Rapaccini and Aylmer in Nathaniel Hawthorne's two short stories 'Rapaccini's Daughter' (1844) and 'The Birthmark' (1846). Wells created his doctor in *The Island of Dr Moreau* (1896) about ten years after Stevenson's. Jekyll shares with his predecessors and successors the same monomania and disastrous ending: all these cautionary tales explore the physical and ethical limits of medicine and science, and show the danger of transgressing them. But Jekyll's story is more markedly tragic, and basically follows the same pattern as ancient Greek drama.

The causes for Jekyll's tragic failure are imperfection and impurity, first taken in the material sense. His confession, and Lanyon's, draw our attention to the imperfect conditions in which his work was carried out. When Lanyon examines the powders that will shortly be mixed and used by Hyde, he points out their home-made appearance:

> The powders were neatly enough made up, but not with the nicety of the dispensing chemist; so that it was plain they were of Jekyll's private manufacture; and when I opened one of the wrappers I found what seemed to me a simple crystalline salt of a white colour. (Stevenson, 1886, p. 55)

In his 'statement', Jekyll has now the benefit of hindsight and laments the incomplete state of his scientific knowledge when he first embarked on his experiment: 'as my narrative will make, alas! too evident, my discoveries were incomplete' (Stevenson, 1886, p. 62). He only realises too late, when his end is near and his fate already sealed, that the drug he used was flawed:

> My provision of the salt, which had never been renewed since the date of the first experiment, began to run low. I sent out for a fresh supply, and mixed the draught; the ebullition followed, and the first change of colour, not the second; I drank it and it was without efficiency. You will learn from Poole how I have had London ransacked; it was in vain; and I am now persuaded that my first supply was impure, and that it was that unknown impurity which lent efficacy to the draught. (Stevenson, 1886, p. 75)

This flaw in the powder is an apt psychological symbol for Jekyll's own tragic flaw. The world of *The Strange Case* is ruled by several types of determinism (social environment and religious background) that the doctor's wrong choices and own personal failings tend to aggravate. Stevenson here explores the interaction between various modalities of fate or relentless determinism and free will. Jekyll's moral hybridity evokes that of the tragic hero as defined by Aristotle's *Poetics* (Aristotle, 4th century BC). He should not be wholly good, otherwise his tragedy would only cause outrage, but he cannot be wholly bad either, otherwise it would be impossible to pity him and experience catharsis. The tragic flaw or *hamartia* leading to his downfall is generally his *hubris* that leads him to ignore or transgress divine laws, *nemesis* always overtaking him. This characterises Jekyll in the very early stages of his experiment at least, and the doctor experiences the same excruciating *anagnorisis* (or recognition) as his Greek forerunners.

Jekyll's brief autobiographical presentation highlights the concept of inheritance, nature and inborn properties or qualities. His wealthy background, bright intelligence and virtuous disposition could therefore have ensured his happiness and prosperity:

> I WAS *born* in the year 18 – *to* a large fortune, *endowed* besides with excellent parts, *inclined by nature* to industry, fond of the respect of the wise and good among my fellow-men, and thus, as might have been supposed, with every guarantee of an honourable and distinguished future. (Stevenson, 1886, p. 60; emphasis added)

The first false note is introduced by the notion of chance, associated with the direction of his scientific studies 'which led wholly toward the mystic and the transcendental' (Stevenson, 1886, p. 60). But knowing the type of claustrophobic religious atmosphere that surrounded the young man, the reader understands there is no such thing as chance in his world, and that seemingly fortuitous circumstances are but disguised manifestations of determinism. As the more mature man is finally and painfully aware in his confession, every attempt to avert one's fate, every seeming chance act, is but an irrevocable step towards one's inescapable doom:

> I have been made to learn that the doom and burthen of our life is bound for ever on man's shoulders, and when the attempt is made to cast it off, it but returns upon us with more unfamiliar and more awful pressure (Stevenson, 1886, pp. 61–2).

As a human being and the inheritor of original sin, he was bound to do wrong by exerting his free will. In *The Disappearance of God*, John Hillis Miller highlights the way rigorous evangelicalism shaped and darkened Thomas De Quincey's conception of existence, but his analysis is quite valid for Jekyll within his Calvinistic background:

> a man's whole life is irrevocably determined by certain crucial acts or choices. A man can never know when he may be standing at one of these crossroads, where a wrong step will lead him into an endless labyrinth of suffering....A sense that at every moment each man hangs over a moral abyss, a feeling that he can never foresee all the results of any act, an experience of anticipatory guilt, since any act is bound to be in some degree evil, an exaggerated sense of scrupulous responsibility – all these evangelical motifs are present in

De Quincey's vision of time as a causal chain, or a labyrinth where one false turn will lose us forever. (Hillis Miller, 1963, p. 60)

That night I had come to the fatal cross-roads. Had I approached my discovery in a more noble spirit, had I risked the experiment while under the empire of generous or pious aspirations, all must have been otherwise, and from these agonies of death and birth, I had come forth an angel instead of a fiend. The drug had no discriminating action; it was neither diabolical nor divine; it but shook the doors of the prisonhouse of my disposition; and...that which stood within ran forth. At that time my virtue slumbered; my evil, kept awake by ambition, was alert and swift to seize the occasion; and the thing that was projected was Edward Hyde. Hence, although I had now two characters as well as two appearances, one was wholly evil, and the other was still the old Henry Jekyll, that incongruous compound of whose reformation and improvement I had already learned to despair. The movement was thus wholly toward the worse. (Stevenson, 1886, p. 64)

Jekyll is both a victim and a sinner, but he judges himself uncompromisingly and does not attempt to disguise his faults, his *hubris* – 'flushed as I was with hope and triumph' (Stevenson, 1886, p. 63), 'my evil, kept awake by ambition' (Stevenson, 1886, p. 64) – and his sinful weakness, his complacency, and his condoning Hyde's misdeeds as something outside himself: 'And thus his conscience slumbered' (Stevenson, 1886, p. 66).[11] The first time he transforms himself spontaneously into Hyde, he knows he has reached another fatal crossroads but only takes half-hearted measures:

Between these two, I now felt I had to choose....I chose the better part and was found wanting in the strength to keep to it.

Yes, I preferred the elderly and discontented doctor, surrounded by friends and cherishing honest hopes; and bade a resolute farewell to the liberty, the comparative youth, the light step, leaping impulses and secret pleasures, that I had enjoyed in the disguise of Hyde. I made this choice perhaps with some unconscious reservation, for I neither gave up the house in Soho, nor destroyed the clothes of Edward Hyde, which still lay ready in my cabinet. (Stevenson, 1886, pp. 68–9)

Symbolically, the effects of the draught decrease as Jekyll's willpower and self-control themselves wane:

> The power of the drug had not been always equally displayed. Once, very early in my career, it had totally failed me; since then I had been obliged on more than one occasion to double, and once, with infinite risk of death, to treble the amount; and these rare uncertainties had cast hitherto the sole shadow on my contentment. Now, however, and in the light of that morning's accident, I was led to remark that whereas, in the beginning, the difficulty had been to throw off the body of Jekyll, it had of late gradually but decidedly transferred itself to the other side. All things therefore seemed to point to this: that I was slowly losing hold of my original and better self, and becoming slowly incorporated with my second and worse. (Stevenson, 1886, p. 68)

Each new lapse from virtue, even if merely mental and not actualised, conjures up Hyde and gives him added strength:

> It was a fine, clear, January day...and the Regent's Park was full of winter chirrupings and sweet with spring odours. I sat in the sun on a bench;...the spiritual side a little drowsed, promising subsequent penitence, but not yet moved to begin. After all, I reflected, I was like my neighbours; and then I smiled, comparing myself with other men, comparing my active goodwill with the lazy cruelty of their neglect. And at the very moment of that vain-glorious thought, a qualm came over me, a horrid nausea and the most deadly shuddering. These passed away, and left me faint; and then as in its turn the faintness subsided, I began to be aware of a change in the temper of my thoughts, a greater boldness, a contempt of danger, a solution of the bonds of obligation. I looked down; my clothes hung formlessly on my shrunken limbs; the hand that lay on my knee was corded and hairy. I was once more Edward Hyde. A moment before I had been safe of all men's respect, wealthy, beloved – the cloth laying for me in the dining-room at home; and now I was the common quarry of mankind, hunted, houseless, a known murderer, thrall to the gallows. (Stevenson, 1886, pp. 71–2)

Hyde's hand is a moral index, a symbol of the power lost by Jekyll that evokes the phrase *to get out of hand*, and, on two occasions, it is the first visible symptom of the doctor's uncontrolled transformations:

> in one of my more wakeful moments, my eyes fell upon my hand. Now the hand of Henry Jekyll (as you have often remarked) was professional in shape and size: it was large, firm, white, and comely.

> But the hand which I now saw, clearly enough, in the yellow light of a mid-London morning, lying half shut on the bed-clothes, was lean, corded, knuckly, of a dusky pallor and thickly shaded with a swart growth of hair. It was the hand of Edward Hyde.
> ... terror woke up in my breast as sudden and startling as the crash of cymbals; and bounding from my bed, I rushed to the mirror. ... Yes, I had gone to bed Henry Jekyll, I had awakened Edward Hyde. (Stevenson, 1886, pp. 66–7)

As in the case of the hand, the meaning of sleep, which soon became Jekyll's worse enemy, is highly symbolic. Goya's 1799 series of eighty aquatints entitled *Los Caprichos*,[12] and more specifically 'Capricho 43' entitled 'The sleep of reason produces monsters' give us the full measure of Hyde's subliminal dimension and power.[13] Jekyll is painfully aware of 'the brute that slept within [him]' (Stevenson, 1886, p. 73), but ironically, it is his own falling asleep that enables the 'beast' to wake up:

> In short, from that day forth it seemed only by a great effort as of gymnastics, and only under the immediate stimulation of the drug, that I was able to wear the countenance of Jekyll. ... above all, if I slept, or even dozed for a moment in my chair, it was always as Hyde that I awakened. ... But when I slept, or when the virtue of the medicine wore off, I would leap almost without transition (for the pangs of transformation grew daily less marked) into the possession of a fancy brimming with images of terror, a soul boiling with causeless hatreds, and a body that seemed not strong enough to contain the raging energies of life. The powers of Hyde seemed to have grown with the sickliness of Jekyll. (Stevenson, 1886, p.74)

For Pascal Nouvel, the gradually irreversible effects of the draught can be seen as an anticipatory description of what is now known as 'addiction':

> Jekyll is somehow becoming addicted to his own draught and this is generally considered as the moral message of the story. Addiction appears to be the price to pay for the exhilarating feeling produced by the drug. Thus, the novella can be read like a disguised personal account of the use of cocaine. (Nouvel, 2010)

But this tale of addiction, if we accept the fact the novella is one and can be regarded as a disguised confession, is told by roundabout means, through suggestion, metaphor and allegory.[14]

The Strange Case of Dr Jekyll and Mr Hyde is a particularly dark and pessimistic text about the quest for transcendence and human enhancement that evinces a very subtle, protoanalytical understanding of the psyche, and exposes the deleterious effects of a repressive social and religious environment. It is also a very powerful latter-day prose tragedy that achieves its cathartic end, terror and pity for Dr Jekyll, who was originally 'more sinned against than sinning' (Shakespeare, *King Lear*, III, 2, 60). Over sixty film adaptations were made between 1901 and 2008 and, except for Stephen Frears' very ambiguous and uncanny *Mary Reilly* (1996) in which the eponymous female servant perceives many things off-stage and from behind closed doors and whose role has sometimes been compared to Utterson's, most of these films defeat this cathartic purpose. They mainly focus on Hyde's physical and moral monstrosity, the atrocities he commits and their affinities with the infamous Ripper murders of the 1880s, whereas his misdeeds are not described explicitly in the novella (except for the murder of Carew), and Jekyll's angst and moral torments are the cornerstone of Stevenson's text. We should remember Jekyll as an unfortunate representative of mankind, and as a man who, when hitting rock bottom, and although at the point of death, still retains his essential humanity, even towards Hyde:

> But his love of life is wonderful; I go further: I, who sicken and freeze at the mere thought of him, when I recall the abjection and passion of this attachment, and when I know how he fears my power to cut him off by suicide, I find it in my heart to pity him. (Stevenson, 1886, p.75)

Acknowledgements

I wish to thank Pascal Nouvel, Professor of Philosophy at the Université Paul Valéry – Montpellier 3, for the valuable scientific and bibliographical information he provided, referred to in the paper as Nouvel (2010).

Notes

1. 'Man is neither angel nor beast and the misfortune is that he who would act like an angel shall act like a beast'. (Translation by F. Dupeyron-Lafay)
2. Indeed, in his *Confessions of an English Opium-Eater* (1821), Thomas De Quincey repeatedly insists on the fact that laudanum did not in the least induce drowsiness but enhanced his mental and artistic faculties. The passage describing his discovery of opium emphasises its rapid exhilarating properties, showing that 'in an hour', he felt 'an upheaving, from its lowest depths, of the inner spirit' and 'an apocalypse of the world' within him, 'an abyss of divine enjoyment' (De Quincey, 1821, 'The Pleasures of Opium', pp. 38–9).

3. James Hogg's *The Private Memoirs and Confessions of a Justified Sinner* (Hogg, 1824) is an interesting Scottish predecessor, a weird, uncanny story of the double, based on a Calvinist pattern of predestination. Hogg's book prefigures the fragmentation and mystery of *The Strange Case*. It features two distinct parts: the first one is a third person objective narrative and the second one is told in the first person by the 'justified sinner' himself – the mentally deranged Robert Wringhim. *The Memoirs* also present several testimonies by other characters.
4. With the quotation from the Book of Daniel (Chapter 5), the reference to Belshazzar's feast prefigures the terrible retribution in store for Jekyll:
    This inexplicable incident, this reversal of my previous experience, seemed, like the Babylonian finger on the wall, to be spelling out the letters of my judgment; and I began to reflect more seriously than ever before on the issues and possibilities of my double existence. (Stevenson, 1886, p. 67)
5. Jekyll mentions 'the perennial war among [his] members' (Stevenson, 1886, p. 60) in his confession.
6. 'There was no mirror, at that date, in my room; that which stands beside me as I write, was brought there later on and for the very purpose of these transformations' (Stevenson, 1886, p. 63). See Paul's 'in a glass darkly' (Corinthians 13:11–12).
7. '... [I]t yet remained to be seen if I had lost my identity beyond redemption and must flee before daylight from *a house that was no longer mine*' (Stevenson, 1886, p. 64, emphasis added).
8. Just as Stevenson denied using drugs similar to Jekyll's potion, Nouvel (2010) shows that, when a journalist from New Zealand interviewed him in 1893, and asked him: 'Had you heard of any actual case of double personality before you wrote your book?', Stevenson responded: 'Never...', but added: 'After the book was published, I heard of the case of Louis V (a case of double personality described by Frederick Myers), but not before'. Nouvel considers that 'Despite this denial and in the light of striking correspondences between Stevenson's work and case studies in French and British popular and scientific journals during the 1870s and 1880s, many critics believe that it is highly unlikely that Stevenson's reply to the reporter was entirely honest. It may be, as Richard Dury speculates, that Stevenson "refuses to collaborate with the reporter because he does not wish to provide a single key to a story that is intended to remain enigmatic"'.
9. The etymology of 'monster' (*monstrare*) refers to showing. Indeed, what the monstrous creatures brought to life by literary scientists such as Frankenstein or Jekyll show is the essence of their own creator. Hyde is a 'monster' demonstrating and embodying the evil potentialities of what is repressed and hidden. Concealment and hypocrisy give birth to monsters.
10. Joseph Sheridan Le Fanu's short story 'Green Tea' (from the collection *In a Glass Darkly*, 1872) features a Protestant clergyman (called Reverend Jennings) haunted and persecuted by a malevolent little monkey – which nobody sees but himself – that finally drives him to suicide. This is more or less what Hyde achieves, but as in Poe's 'William Wilson', the death of one means the death of both.
11. 'I sat in the sun on a bench; the animal within me licking the chops of memory; the spiritual side a little drowsed' (Stevenson, 1886, p. 71).

12. The reader can consult the full series in a pdf file at the address: http://www.gasl.org/refbib/Goya_Caprichos.pdf (last consulted: July 2010).
13. Jekyll seems to be dimly aware of this when he refers to his alter ego as 'inorganic' and as an 'amorphous dust' (Stevenson, 1886, p. 73).
14. Nouvel evokes Wilde's (negative) opinion on *The Strange Case*: 'the transformation of Dr Jekyll reads dangerously like an experiment out of *The Lancet*'. This sounds grossly unfair in view of the very scarcity and vagueness of medical elements, evoked in the second part of this paper, and the very Gothic mood that prevails whenever the laboratory or the experiments are dealt with.

References

Aristotle (4th century BC) *Poetics*, Sachs J. (tr.) (London: Penguin Classics 'The Focus Philosophical Library', 2006).
De Quincey T. (1821) *Confessions of an English Opium-Eater*, Lindop G. (ed.) (Oxford: Oxford University Press 'Oxford World's Classics', 1998).
Frears S. (1996) *Mary Reilly* (film) (USA: TriStar Pictures).
Freud S. (1916–17) *A General Introduction to Psychoanalysis* (Ann Arbor: University of Michigan Library, 2009).
Freud S. (1884) 'Über Coca', in *Centralblatt für die Gesamte Therapie*, Vol. II (Vienna: W. Stein for Moritz Perles, 1885), pp. 289–314, http://fr.scribd.com/doc/68768839/Uber-Coca-Pt-1.
Goya F. (1799) *Los Caprichos* (print) (Middletown CT: Wesleyan University, Davison Art Center), http://www.wesleyan.edu/dac/coll/grps/goya/goya_intro.html (accessed July 2010).
Hawthorne N. (1844) 'Rapaccini's Daughter', in Hawthorne N. (1846) *Mosses from an Old Manse* (Las Vegas NV: IAP, 2009).
Hawthorne N. (1846) 'The Birthmark', in Hawthorne N. (1846) *Mosses from an Old Manse* (Las Vegas NV: IAP, 2009).
Hillis Miller J. (1963) *The Disappearance of God. Five Nineteenth-Century Writers* (Cambridge MA: The Belknap Press of Harvard University Press).
Hogg J. (1824) *The Private Memoirs and Confessions of a Justified Sinner*, Carey J. (ed.) (Oxford: Oxford University Press 'The World's Classics', 1995).
Le Fanu J.S. (1872) 'Green Tea', in Tracy R. (ed.) (1993) *In a Glass Darkly* (Oxford: Oxford University Press 'The World's Classics').
Nouvel P. (2010) 'Unpublished Commentary, October 2010', in *Human Enhancement: an interdisciplinary inquiry*, workshop #5 [Paris: CERSES (Université Paris Descartes, CNRS) and IHPST (Université Panthéon-Sorbonne Paris 1, ENS, CNRS), 2010].
Pascal B. (1670) *Pensées*, Le Guern M. (ed.) (Paris: Gallimard 'Folio classique', 1977).
Poe E.A. (1839) 'William Wilson', in Colum P. (ed.) (1981) *Tales of Mystery and Imagination* (London: Dent 'Everyman's Library').
Rank O. (1914) *The Double: a Psychoanalytic Study*, Tucker H. Jr. (tr.) (Chapel Hill: University of North Carolina Press, 1971).
Schultz M. (1971) 'The Strange Case of Robert Louis Stevenson: a Tale of Toxicology', *Journal of the American Medical Association*, 216(1), 90–4.

Shakespeare W. (c. 1605) *King Lear*, in Alexander P. (1951) *The Complete Works of William Shakespeare* (London and Glasgow: Collins, 1983).
Shelley M. (1818) *Frankenstein*, in Fairclough P. (ed.) (1983) *Three Gothic Novels: The Castle of Otranto, Vathek, Frankenstein* (Harmondsworth: The Penguin English Library).
Stevenson R.L. (1884) 'Markheim', in Stoneley P. (ed.) (1991) *The Collected Shorter Fiction* (London: Robinson Publishing).
Stevenson R.L. (1886) *The Strange Case of Dr Jekyll and Mr Hyde*, Letley E. (ed.) (Oxford: Oxford University Press 'Oxford World's Classics', 1998).
The Bible. Authorized King James Version, Carroll R. and Prickett S. (eds.) (Oxford: Oxford University Press 'Oxford World's Classics', 2008).
The RLS Website, http://www.robert-louis-stevenson.org/richard-dury-archive/films-rls-jekyll-hyde.html. (accessed July 2010).
Wells H.G. (1896) *The Island of Dr Moreau*, in Parrinder P., Atwood M. and McLean S. (eds) (London: Penguin Classics, 2005).

11
Biotechnology and the Future of Sport: A Scenario
Jean-Noël Missa

Introduction

I recently discovered an unpublished manuscript in the library of my university, a text that studies the evolution of sport in the 21st and the early years of the 22nd century. This manuscript of unknown authorship was probably written by a philosopher and historian of sport just after the Olympic Games held in Brussels in 2144. As an introduction to a detailed description of this manuscript from the future, I will begin with a brief analysis of the current situation of enhancement in sport.

Enhancement and sport at the beginning of the 21st century

The renewal of the debate on performance enhancement in sport

The debate on performance enhancement has been considerably renewed in the last decade. There are two reasons for this:

The creation of the World Anti-Doping Agency (WADA)

The first reason is the establishment of new regulatory structures and standards for enhancement in sport. After a period of relative tolerance towards doping in the 1980s and the early 1990s, the World Anti-Doping Agency was created with the aim of strengthening the fight against doping. In the 1998 Tour de France, the Festina Team was excluded from competition after a member of the coaching staff was found with a car full of drugs. Other teams withdrew in protest, some of whom were themselves under suspicion for cheating. This doping scandal led to the creation of the World Anti-Doping Agency (WADA). WADA was established in 1999 as an international independent agency funded by the sport movement and the governments of the world. WADA works towards a vision of the world that values and fosters a doping-free culture in sport; it has thus developed a war-like ideology similar to the political war against illicit drugs. WADA's role is to promote, coordinate and monitor the fight against drugs in sport. Current anti-doping strategy aims to eradicate doping in elite sports by means of repression. Some philosophers think that eradication of doping in sport is an unattainable goal. They favour a more pragmatic approach that allows doping under medical supervision.

The emergence of enhancement medicine

> Dr Dumas, is it possible to turn an average athlete into a great champion simply by using drugs?
>
> –The question is quite difficult to answer. It is a fact that we are currently able to create supermen, just as one is able, by scientific means, to walk in space and perhaps one day land on the moon. It is possible – I don't really know, I am not able to answer your question – that we could turn a man of average value into a superstar. But be that as it may, this is an extremely important moral problem, for where does one draw the line? Will we eventually medicate lathe turners at Renault to increase their performance? Or try to turn an average eight-year old child into a graduate of the École Polytechnique? Will we take a ten year-old child and decide that he should become a great swimming or cycling champion and, from that moment on, use all the means of science without asking ourselves what might happen later, to transform him into a superman at 21 years of age? Laboratory champions? We French doctors, I believe, don't agree with this. (in Chapatte and Gavinet, 1962)

The second reason that gives doping a deeper philosophical and ethical dimension is the emergence of a new medical paradigm: enhancement medicine. The question of enhancing performance in sport has become part of a broader societal debate on human enhancement. The gradual blurring of the boundary between therapeutic medicine and enhancement medicine constitutes the most spectacular and the most troublesome form of these modifications. In contemporary biomedicine, novel medication and technologies can be used not only to cure patients but also to enhance human capacities: modify genetic design, alter cognitive and emotional function, increase lifespan or boost performance in sport, and so on. This development represents a paradigmatic change in medical practice: expectations are no longer limited to the mere restoration of health or the promotion of health. What are required are the improvement of performance and the perfectibility of the human being, including in the field of sport. Competitive sport could become the main laboratory of enhancement medicine. Athletes are risk takers. Coaches and athletes are ready to take experimental drugs and try out new technologies to improve their performance. In order to beat records and win medals, they agree to be the guinea pigs of a huge and enduring clandestine experimentation in enhancement biotechnology.

The intersection of science and sport raises fundamental philosophical, ethical and policy issues that cannot easily be resolved. A war on doping is not the only solution. The ethical approach supporting the prohibitionist attitude is not the only one available. Nowadays, some physicians and philosophers argue that we would be better off if we legalised some forms of enhancement technology in sport under medical supervision. Their arguments deserve to be taken seriously. We need to open a broader and open-minded societal conversation on this topic.

Biotechnology, enhancement and sport: the example of gene therapy

> What is clear ... is just how impatient some coaches and athletes are to find new and ingenious ways to cheat. First it was steroids, then EPO [erythropoietin], then human growth hormone – and now the illicit grail seems to be gene therapy. (Friedmann et al., 2010)

In recent decades, doping in sport has been enabled by advances in pharmacology (amphetamines, steroids, growth hormone, EPO, and so on). More recently, the development of gene therapy has provided new ways for enhancing performance in sport. The blurring of the boundary between therapeutic medicine and enhancement medicine is perfectly illustrated by the example of gene therapy in sport. Today, gene therapy offers tools that make it possible to genetically modify functions affecting normal human traits, including athletic performance. Advances in gene technology may well not only mitigate the effects of diseases such as muscular dystrophy, but also restore the strength of senior citizens and boost performance in sports. Researchers have discovered dozens of genes that affect athletic performance. Scientists have created mice whose bodies are flooded with oxygen-carrying red blood cells, creating greater endurance. Other mice have been engineered to metabolise fat and carbohydrates in such a way that they can run like marathoners without training. This experiment on mice depends on small molecule modulators of peroxisome proliferator-activated receptor-delta

(PPAR-delta modulators) that regulate the expression of genes involved in lipid metabolism and energy utilisation, and increase the production of slow twitch oxidative energy-efficient muscle fibres. These effects have important implications for the therapy of obesity and muscle disease. But this experiment could also have implications for enhancement in sport: mice overexpressing a PPAR-transgene show enhanced endurance performance. Other genetic methods have been used to demonstrate enhanced muscle function from insulin-like growth factor (IGF-1) or follistatin transgenes. Gene therapy can also increase erythropoietin-enhanced blood production in primates, enhancing endurance performance through blood.

One of the first experiments with consequences for enhancing performance in sport was undertaken in 1995 by Se-Jin Lee, a professor of molecular biology at Johns Hopkins Medical School in Baltimore. Lee identified the function of myostatin, a protein that tells muscles when to stop growing. Experimenting on mice, Lee inactivated both copies of the gene in the animal that codes for myostatin. As a result, the rodents developed twice their normal muscle mass. After Lee's results were published in 1997, he received emails from people with muscle-wasting disease, but also from athletes and bodybuilders offering themselves as subjects for human experimentation (McPherron et al., 1997).

In 1998, H. Lee Sweeney, a physiology professor at the University of Pennsylvania, published a paper with a detailed description of the mice that he had injected with a gene engineered to produce a muscle builder called insulin-like growth factor (IGF-1). Sweeney was also inundated with inquiries from athletes. He says a high school football coach and a high school wrestling coach volunteered their entire teams as guinea pigs. 'Even when I tell them it's not safe', Sweeney said, 'some athletes are willing to try anything' (Wenner, 2008).

If athletes could use such gene therapy to block the expression of the gene that codes for myostatin or increase the production of IGF-1, the alteration would be integrated into the DNA of the cells targeted by the procedure. The simplest way of detecting the new gene would be to remove a piece of muscle and probe for it, a procedure most likely too invasive for wide-scale use.

Since the gene genie escaped from the bottle a decade ago, researchers have discovered dozens of genes that appear to affect athletic performance. These scientific approaches are well-known in sport communities, and some of these genes may be used by athletes to enhance performance. Some attempts to use gene technology in sport have already been made. A German athletic coach was found attempting to obtain

Repoxygen, a gene-transfer vector that induces expression of the erythropoietin gene. A Chinese genetics laboratory reportedly offered gene-based manipulations before the 2008 Olympic Games in Beijing. It is not clear whether these or other similar attempts reached the stage of actual use in human athletes, but there seem to be few technical barriers standing in the way. 'One delivery method – flushing the bloodstream with the desired gene – is simple enough', states Sweeney, 'that it could be achieved by a clever undergrad in a molecular biology lab' (Wenner, 2008). The day gene enhancement technology becomes a reality in sport, it is an understatement to say it will be hard to detect the 'cheats'. But 'doping' is not necessarily cheating. Whether it is or not depends on which philosophy of sport you espouse.

Philosophy of sport

WADA has developed a war-like ideology similar to the political war against illicit drugs. There is no evidence that it is the best attitude to have from an ethical point of view. There should be an open discussion of the ethical and philosophical foundations of anti-doping policy. Some philosophers think that the eradication of doping in sport is an unattainable goal (See Kayser et al., 2005, S21; Miah, 2005; Savulescu and Foddy, 2007, 2009). They support a more pragmatic approach that allows doping under medical supervision. We agree with Kayser, Mauron and Miah when they say that 'the ethical foundations of sport are also a matter of public debate and, like for other ethical policies in society, there should be mechanisms ensuring accountability of policy to the broader public' (Kayser et al., 2007). The legitimacy of the war on doping should become a public issue. Let us examine the pros and cons of these two ethical and philosophical approaches.

We have identified six central arguments that are often neglected in the debate on doping.

First argument – Competitive sport is not egalitarian

> Running fast is a gift, something you are born with. No matter how hard you train or whether you have the greatest coach and back-up in the world, if you aren't made to run fast, it isn't ever going to happen. (Chambers, 2009, p. 5)

In WADA's philosophy, fairness is based on the respect of natural inequalities. The reality is that competitive sport is not egalitarian. Basically, the winning athletes are those who have the best genetic predispositions and the best training and medical environment. In the 'war on doping' philosophy, a 'level playing field' is a field where the athlete with the best natural capacities and the best environment to maximise them is going to win. Equality has nothing to do with competitive sport. Professional sport only rewards biological and artificial inequalities. WADA and the sport zealots want to promote a pure sport with athletes taking only water and authorised substances. WADA zealots claim that 'doping is ruining sport, doping is cheating'. Some philosophers disagree with this point of view. They argue that the ethical foundation of the war on doping 'consists of largely unsubstantiated assumptions about fairness in sports and the concept of a "level playing field"' (Kayser et al., 2007). Kayser, Mauron and Miah think that WADA's ethical foundation relies 'on dubious claims about the protection of an athlete's health and the value of the essentialist view that sports achievements reflect natural capacities' (Kayser et al., 2007).

Second argument – Doping is a logical consequence of the nature of competitive sport: 'maximising performance'

> Everybody wants an edge, everybody wants to win. That's the way it is. That is the sport. (Ben Johnson, in Belton, 2001)

The prohibition of doping introduces a structural contradiction within competitive sport. Doping is nothing but the logical consequence of the quest to maximise performance in sport. The athlete is asked to perform, to surpass themselves; but, at the same time, they are required to abstain from using the biotechnological resources that allow performance enhancement. Nobody is going to break Florence Griffith-Joyner's 1988 10'54 100 m record, Marita Koch's 1985 47'60 400 m record, Jamila Kratochvilova's 1983 1:53'28 800 m record (the longest standing individual world record in Track and Field) or Marco Pantani's 1997 37'35 record ascent of the Alpe d'Huez, without using performance enhancing drugs or technology. Some records are impossible to break with a 'natural body'.

Third argument – Doping is part of the reality, the spirit and the history of sport

> I firmly believed that I was the only athlete in the world not cheating. (Chambers, 2009, p. 100)

Whether you like it or not, you have to admit that doping is part of the reality, the spirit and the history of sport. WADA claims that performance enhancing drugs are against the 'spirit of sport'. History shows that this is not true. In cycling or in athletics, for example, taking drugs has always been tacitly part of the rules of the game. Some of the greatest cyclists – Coppi, Kübler, Anquetil, Mallejac, Simpson, Fignon, Pantani – have admitted to using drugs. Everyone knows that most professional cyclists need to take drugs if they want to be competitive. Interviewed by Oprah Winfrey in 2013, Armstrong said that in order to keep on winning the Tour de France, an athlete has to keep on using banned substances to do it: 'That's like saying we have to have air in our tyres or we have to have water in our bottles. That was, in my view, part of the job'.[1] It is also true for other sports.

Doping is part of the essence of competitive sport. The nature of professional sport forces the athletes to complete their training with a biomedical preparation. We can regret this fact and live in the nostalgia of a 'pure sport' that has never existed. Nevertheless, it is a fact that biomedical technology is at the heart of the performance-enhancing philosophy of elite sport. Should we not consider as highly paradoxical the determination to outlaw a behaviour that is at the core of competitive sport: enhancing performances through artificial means? The pragmatic attitude advocated by Kayser, Mauron and Miah seems more adequate for professional sport:

> Elite athletes are also constituted by scientific knowledge and this is a valued aspect of contemporary sport. As such, translating doping enhancements into earned advantages – having the best scientists on one's team – would more closely align to the values of competition than leaving it all to chance, unequal access to illicit practices, and the cleverness of undetected cheating. (Kayser et al., 2007)

Fourth argument – Anti-doping philosophy is a source of prudishness and hypocrisy in sport

> They are cheating you, Dwain. You're a very talented athlete but you are not competing on a level playing field. The system allows people to cheat. (Victor Conte, in Chambers, 2009, p. 61)

Since WADA's creation, there has been a new prudishness about doping that is almost unbearable. Everyone knows that one has to take drugs to win in certain sports, and yet everyone seems astonished and morally shocked when an athlete is controlled positive. The athletes have to take drugs in order to win but they must not be caught testing positive. They are confronted with two systems of rules: the official rules according to which it is a sin to take drugs; the unofficial rules telling them there is no other way they can win the competition. The athlete has to take drugs and, at the same time, he must pretend he is against doping. He must 'cheat' and pretend he is not a 'cheat'. The discrepancy between the two systems of rules represents a terrible burden for the athlete. The 'anti-doping war' creates a very strange situation for elite athletes that increases the complexity of their already difficult job. Legalising drugs would at least do away with the hypocrisy about doping in sport.

Fifth argument – The boundaries between authorised and unauthorised doping are constantly changing and arbitrarily delineated

What is the difference between Colette Besson, Lasse Viren, Kenyan and Ethiopian long-distance runners, Eero Mäntyranta, Bjarn Riis, Marion Jones, Marco Pantani, Riccardo Ricco, and Floyd Landis? Besson, Viren, Mäntyranta, and the long-distance African runners are all in the green authorised zone. Colette Besson, the former French Olympic 400 m champion in 1968, was one of the first athletes to train in altitude to increase her haematocrit level. The Ethiopian long-distance runners (Kenenisa Bekele, Haile Gebreselassie, and so on) live naturally in a favourable environment. It is has been said that Lasse Viren, former Finnish long-distance runner, winner of four gold medals at the 1972 and 1976 Olympics, won his medals with the help of 'blood doping'. But this technique was not forbidden by the International Olympic Committee (IOC) in the 1970s. Eero Mäntyranta is a 'naturally doped' athlete, thanks to a genetic mutation that induces a change in the erythropoietin receptor, increasing the sensitivity of erythroïd progenitor cells, and leading to a high haematocrit level. Mäntyranta, whose blood carries more haemoglobin and therefore more oxygen than ordinary people – a large advantage when participating in endurance events – won three gold medals in cross-country skiing at the 1960 and 1964 Winter Olympics. Besson, Viren, Bekele, Gebreselassie and Mäntyranta were not condemned to a partial or lifetime ban on professional competition. Their 'doping' was natural or tolerated by the rules of sport. Riis took erythropoietin without being caught in the 1996 Tour de France.

Jones, Pantani, Ricco and Landis were not so lucky. Having used erythropoietin or other substances to improve their performance (as did most of the athletes against whom they were competing), they were condemned for doping, and their life was partially or totally destroyed. Marco Pantani, for example, an Italian road racing cyclist, one of the most gifted climbers of the 20th century, was disqualified from the 1999 Giro because his haematocrit level was above the limit of 50. Morally defeated by doping suspicions and persecutions, he suffered clinical depression and died alone in a hotel room in Rimini in February 2004. Marion Jones, one of the greatest female track and field athletes of all times, was sentenced to six months in jail because she lied when she denied using steroids before two grand juries. Her sentence was an indirect consequence of the war-on-doping policy. Floyd Landis, a former teammate of the seven time Tour de France winner Lance Armstrong, was stripped of his title in the 2006 Tour de France because of a doping offence. He was suspended from professional competition through 2009. Bjarn Riis was luckier in escaping a similar doping offence. Nicknamed 'Mr 60%', a suggestion that he used erythropoietin to win the 1996 Tour de France, Riis confessed to taking EPO, growth hormone, and cortisone for five years from 1993 to 1998, including in 1996, the year of his victory on the Tour de France. Riis was first removed from the official record books of the Tour de France, but in July 2008 he was written back into the books, along with additional notes about his use of doping. Landis has disappeared from the record books, but not Riis. Why?

These examples reveal the blurred, artificial border between legal and illicit performances, between natural and artificial doping, between winning a gold medal and being despised as a cheat. Why punish and destroy the life of Landis, Pantani, Ricco, Marion Jones and glorify the others? Four of the top five finalists at the Seoul 1988 Olympic 100 m final tested positive for banned drugs at some point in their careers. Somewhat unfairly, of the four, only Johnson was forced to give up his medal. This is real unfairness. The current rules should be changed.

Sixth argument – The 'anti-doping' policy is ineffective and has negative consequences

Ineffectiveness

> The war on doping can never be won. In doping, you can only get partial victories. (Samaranch cited in *New York Times*, 2001)

Doping is a fact. A 'zero tolerance strategy' is bound to fail. Drugs such as EPO or growth hormone are hard to detect. Cheats will always find new drugs or new technologies (gene doping, stem cell injections, and so on) that are almost impossible to detect. The battle against drugs can never be won.

Threat to privacy

The war on drugs in sport is a threat to privacy. 'Testers' can turn up whenever they like, before, during or after a competition, and any other time during the year. If an athlete fails to meet them on three separate occasions in an 18-month period, they could incur a ban. We should stop treating elite athletes as potential criminals who have to inform the authorities of their every move. The world's number one tennis player Rafael Nadal has attacked the drug testing procedure, saying 'he feels like a criminal' (Rafael Nadal, BBC Sport, 12 February 2009). He is angry about how much information he has to produce about his whereabouts. He has to say where he will be for at least an hour every day, seven days a week. The privacy of athletes should be protected. In Belgium, 65 sportsmen have launched a legal battle against WADA, claiming that they are intrusive and that they infringe European privacy laws.

Criminalising sport and demonising the athletes

> It is amazing, you know. A triple murderer hasn't had the kind of criticism that he got. (Charlie Francis, Ben Johnson's coach, in Belton, 2001)

In theory, WADA policy promotes fair competition. In reality, this is not the case. The history of sport in the last decade has shown that the zero tolerance policy has failed. Because present doping controls are ineffective, they put honest athletes who do not take drugs at a disadvantage. Because of the relative inefficacy of these controls, most athletes know that they must take drugs in order to win. The cleverest and the luckiest 'cheats' are rewarded. The others, unlucky enough to be controlled positive, will have their career and life broken by a two-year, four-year or even a lifetime ban. We should not demonise great athletes such as Marion Jones, Lance Armstrong or Barry Bonds. We should leave them alone. The death of the Italian cyclist Marco Pantani could be interpreted as a consequence of the zero tolerance policy.

Health risks caused by clandestine doping
In theory, WADA policy protects health. This is not necessarily the case. Athletes are under pressure to find undetectable enhancers with no special attention to safety. Drugs are often produced on the black market and administered in a clandestine, uncontrolled way. Only the wealthiest athletes can hire a private physician to advise them. Most of the athletes receive drugs through their coaches or drug dealers, neither of whom have medical training.

Continual rewriting of sport history
Who won the track and field women's 100 m final at the 2000 Sydney Olympic Games? Marion Jones? Yes and no. In 2007, the IOC formally stripped Marion Jones of her medals. Though the International Association of Athletics Federations (IAAF) lists Ekaterini Thanou as the first-place finisher in the women's 100 m race, she was not awarded a gold medal by the IOC. So the women's 100 m Olympic race in Sydney is a race without a winner. The same thing happened at the 2006 Tour de France: in early August 2006, Landis was found guilty of doping and was disqualified. The second place rider, Oscar Pereiro, became the official winner of the race. But you will never find a cycling fan who is ready to accept the fact that Pereiro won the Tour de France. The 2006 Tour de France is a race without a winner. Landis accused several former teammates, including Lance Armstrong, of using EPO and blood transfusions in the 2002 and the 2003 seasons. Armstrong has since become the new victim of the witch-hunt. And the process will go on beyond Armstrong[2]. The sports authorities are constantly reviewing the books to be sure that the true 'Spirit of sport' is the real winner of every competition. A close examination of the 'doping offence' files concerning former cyclists or track and field Olympic athletes could transform the Tour de France and the Olympic Games winner lists into blank sheets. If this re-evaluation of sport history continues, the WADA and IOC historians may one day conclude that no one has ever really won a Tour de France, or a track and field Olympic final, in accordance with the true 'Spirit of sport'.

The inevitability of a biotechnological evolution of sport

It is difficult to decide which ethics best accompanies the harmonious development of sport. There is no easy solution to the problem of doping. But a pragmatic approach that allows some forms of performance enhancement under medical supervision is certainly more consistent with the overall philosophy of competitive sport. So, the anti-doping ideology could well lose ground in the near future because it is at odds with the reality of competitive sport. The anti-doping ideology will probably share the fate of the 'pro-amateurism and anti-professional' ideology that died in the 1970s when it came into complete contradiction with the development of competitive sport. Beyond the debate over the 'war on doping', we have the feeling that some forms of biotechnological enhancement in sport cannot be prevented. This is also the opinion of Ted Friedmann, a specialist of gene therapy in sport medicine:

So why does one think that genetic approaches to athletic enhancement are inevitable? First of all, athletes are risk-takers. They're young healthy athletes who think nothing is ever going to happen to them. And they are known to accept all sorts of risks. Polls have been taken of young athletes asking if I were to guarantee you a gold medal in the next Olympics at the risk of losing 20 years of your life would you do it? And universally, they say yes. They will take that risk for the reward of gold medals. There are enormous financial pressures and national pressures to push athletes to perform and to win. We know that they use pharmacological approaches to enhancement. We know that they're aware of gene transfer technology, and we know that that technology is still immature, but it's advancing rapidly. And we know that many of the studies in gene transfer technology, in fact, use the genes that are of particular interest to athletes, Erythropoietin, growth hormones and other relevant genes. ... enormous pressures exist in athletics which make this kind of direction very likely, and inevitable. (Friedmann, 2002)

Whether we like it or not, the most probable scenario for the future of sport includes a necessary increase in the use of medical knowledge and technology to improve performance. The mysterious manuscript I promised to present seems to confirm the idea of an inevitable biotechnological evolution of sport.

An unpublished manuscript on performance enhancement in sport

It was on a dusty set of shelves of the University of Brussels' library that I discovered, last autumn, a black folder containing the manuscript I am going to describe. It appears to be the detailed transcript of a presentation that was given – or more precisely, that will be given, as the timeline is blurred – at the bicentenary celebration of the creation of the University of Brussels' Institute for the Philosophy of Sport.

The file contains an appendix mentioning some meagre historical facts. Some historians of the future, with whom I have shared the manuscript, have voiced the hypothesis that the text could be the inaugural conference of a prestigious international colloquium held in October 2145, less than a year after the Brussels Olympic Games. The text was not signed but was very probably written by a Belgian specialist in the field of biomedical human enhancement. Conventionally, we will name him PKD, as a tribute to the great American science fiction writer.

Let us now consider the text itself, which I wish to present as soberly as possible. The central theme of PKD's inaugural conference was the enhancement of human physical and mental performance. The text lays out a historical study of the theme, as our colleague of the future looks back on the genesis and development of human enhancement technologies, and the vivid ethical questions they inevitably raised during the 21st century. After the 2144 Brussels Olympics held its closing event, featuring eight athletes representing the eight powerful biotechnological firms, PKD decided to tackle the question of performance enhancement in sport. He approached it as a paradigmatic illustration of a more general reflection, historical as well as ethical, on enhancement medicine and the perfectibility of the human being.

In analysing this material, I will follow PKD's presentation. The study, divided into four quarters, is structured as a Rollerball game. To avoid frightening his audience, PKD does not allude to the length of the quarters but specifies that, in its latest version, Rollerball is being played without a time limit. The occurrence of extra-time after the end of the fourth quarter is unfortunately not a rare case, and depends on the progress of the game. PKD promises, however, that everything should be over before midnight. But night has not yet fallen, far from it, when PKD gives the starting signal for the first quarter, entitled 'Brussels 2144'.

Brussels 2144

In the wake of the 2144 Brussels Olympics, the question of performance enhancement in sport began to preoccupy PKD. He was extremely enthusiastic about these games, especially the 100 m final, the supreme competition in track and field athletics. The narrator had been present in the Raymond Goethals Stadium when the eight athletes were introduced to the crowd. The new structure with an impressive 250,000 attendance capacity, a holy temple erected to the sacred glory of the Olympics and the emerging religion of sport, had been specifically designed for the Olympics on the site of the previous Basilica of Koekelberg. The remains of Saint Raymond, apostle of the new cult of sport, were carefully preserved as a precious relic in the walls of the Stadium. Among the contestants for the 100 m race, four front-runners were breaking away. In lane number three, there was quite naturally Parker, the Jamaican gold medallist from the New Delhi Olympic Games and proud world record holder, running for Nike Biotech. A year ago he was still unbeaten. But with the rise of the American teenage prodigy, Carl Jones, who was only 19 years old, the race was still wide open. The young runner – holder of the world's best time in that year attained in Atlanta – had recently been bought by New Pharmaceutics, and followed in lane number four. Next to him, highly focused and warming up, Li Ping, the Chinese rising star from Shanghai Transgenium and the jewel of Asian biotechnology. Last but not least, Jeff Koens, the Belgian sprinter representing European research.

Unless some extremely unexpected event occurred, the other four competitors were destined to struggle for the fifth place. Despite their determination and courage, the technological gap between them and the favourites was almost impossible to fill. The starter's handgun rang out loudly on the track of the covered stadium, in which the atmosphere was frequently adapted to maximise performance. For the sprint races, the 'techno-climate control system' had been set for dried heat with a rarefied oxygen atmosphere, close to the conditions of the 1968 Mexico Games that had allowed many records to be broken in sprint races and field events, including Bob Beamon's spectacular long jump of 8.90 m, a record that it took 22 years to beat.

Such longevity was, without a doubt, unimaginable in PKD's time. The pressure of the audience was far too great. On the track, the athletes responded to the starting signal. Parker and Jones were the first two out of the starting blocks. Their reaction time was of an extraordinary velocity. They looked like springs made of flesh. Needless to say, both Nike Biotech and New Pharmaceutics entertained a reputation for the high quality of their attention modulators. Li Ping had been slower on the starting blocks. Asian science was still catching up in the field of cognitive enhancement. Launched ferociously on the track, the eight athletes were loosening up their muscles to express the full potential of their power and suppleness, both acquired in the strenuous pain of intensive training. Stereotyped gestures, endlessly repeated during the training sessions, allowed the athletes to propel themselves across the finishing line in less than 8 seconds.

'What a magnificent show', PKD was thinking, particularly because he was sharing this moment with his son, Tom, sitting next to him in the terraces. He was experiencing some difficulty in breathing because the rarefaction of oxygen in the air had amplified a slight asthmatic bronchitis.

That very morning, in front of the entrance gates, PKD had been able to get tickets at an exorbitant price through the black market. He wished to initiate his son to the beauty of the Games. He had spent a small fortune for a race that lasted merely a handful of seconds. If anything, it was pure madness! But what entertainment! Tom's eyes were fixed on the athletes. After 60 m on the black cinder track, Jones, the feline, and Parker, the muscles machine, were still even. But 20 m from the finish, Li Ping's profile was getting dangerously close. The commentator's voice was echoing throughout the stadium: 'Parker's still ahead. It's the final battle now, between Parker and Jones. Parker... Parker... Jones is now slightly left behind. He's trying to resist the Jamaican's dominating stride. But now... in lane number five, Ping is making a surge to the finish. Parker remains in front... Li Ping's coming up from behind... Li Ping... Li Ping is Olympic champion!'

Thanks to his stunning final effort, the Chinese runner had crossed the finish line in first position, in front of Parker and Jones. Koens had to be content with fourth. The Chinese victor had achieved another gold medal for Shanghai Transgenium, allowing his firm to surpass Nike Biotech in the overall ranking of the Games. The Chinese journalists rushed to interview the winner. Li Ping, with extreme oriental sobriety, declared himself to be moved and satisfied, thanking his parents, his coach and of course, the entire scientific team of Shanghai Transgenium.

With a time of 7.84 seconds, he had absolutely crushed the world record. The traditional blood and urine samples were evidently collected. They were simply considered as routine checks that the Association for the Enhancement of Performance (AEP) was imposing, not to detect a fraud – the naturalist ideology had been abandoned a long time ago – but to evaluate and ultimately better control the risks incurred by the athletes. It was thus noted that Li Ping's ratio of testosterone/epitestosterone was 12/1. A century before, such a rate would have sparked indignation from the public and faked indignation from sports' journalists. Our colleague of the future is recalling with a smile the time when exceptionally gifted athletes such as Dwain Chambers, Ben Johnson, Tim Montgomery, Marion Jones, Justin Gatlin, and many more, had been sanctioned for using products or techniques to enhance their performance – for doping, the expression which had been established in the 20th century. Back in those days, athletes were disqualified when tested positive. Disqualified! The word had just fallen from use, and PKD evoked with mild sympathy and an obvious lack of understanding that peculiar time when an attempt was made to defend a romantic and naturalistic ideal of sport: the perspective of the athlete winning a race, using nothing but clean water and the cultivation of natural talent through hard work, strong will and courage. Undoubtedly, for PKD, this was behaviour from a distant past, hard to grasp for a 22nd century man.

This is not to say that, for the athletes of the 2144 Brussels Olympic Games, hard work, strong will and courage were not necessary. They were essential. But they were reduced to very little without the technology of the body and mind, an inexhaustible source of new successes, medals, records, and profits...biotechnological genius was reshaping the athlete's body to prepare it for new feats.

In his report, our colleague PKD was looking back on the origins of the modern forms of sports competition, at a time when we were still using the word 'doping' and assimilating it to cheating. 'We need to stop the cheaters', Dick Pound, the president of WADA, the World Anti-Doping Agency, had proclaimed at the turn of the 21st century. That organisation had been replaced in 2024 by the AEP, the Association for the Enhancement of Performance. Was man not born to beat records? And to achieve this goal, was it not mandatory to recombine, enhance and stimulate the athlete's body in order for him to respect the Olympic ideal: 'higher, stronger, and faster'?

PKD studied the roots of this axiological change, of this inexorable transformation of the status of doping: a practice that was first slandered, abhorred and fought against, before being accepted as an inevitable evil, and then, finally encouraged and judged as an intrinsic characteristic of high-level professional sport. Technological genius was used to enhance the athletic act for the greatest pleasure of the spectator. After all, on what grounds could a society that encourages the enhancement of performance in all other sectors of life, ban techniques that constantly improve athletic results and entertain a crowd eager for a constantly renewed improvement of these results? According to PKD, an important step in the axiological change leading to the acceptance of biotechnological enhancement took place in New York, during the 2024 Summer Olympic Games, more precisely the day Antonio Perez, chairman of the IOC, gave his inaugural speech, a discourse on the philosophy of competition that deeply influenced the future of sport.

IOC President Antonio Perez's inaugural speech and the end of prohibition

For the opening ceremony of the New York 2024 Olympics, an immense crowd had gathered on the banks of the East River to admire the sailboat procession between the Olympic stadium and the village. The crowd watching the athletes' parade on Broadway before entering the new sports building was just as numerous.

New York was offering a concept of the games centred in the inner part of the city. The facilities were located alongside two key transportation routes. The North–South route ran along the East River and crossed the East–West route, represented by the horizontal axis of 47th street, thus drawing an Olympic X that symbolised the abolition of prohibition and the new alliance of Man and Technology. The Olympic village was situated exactly at the intersection and had been built in the Queens district, at the foot of the Queensborough Bridge. The bridge thus physically and symbolically linked the village to the Olympic stadium, recently inaugurated on the banks of the Hudson River, on the west side of Manhattan and named, without a great deal of originality, Freedom Stadium. Possibly, the name referred to the athletes' new freedom to enhance their performances.

After this brief description of the different locations hosting the New York City Olympic Games, PKD focused his attention on Antonio Perez's inaugural speech. The Mexican lawyer had been elected IOC President in 2021. For several years now, Perez had been struggling to stop the hypocritical carnival in the sporting world concerning doping. Perez's joy was clearly perceptible in his tone of voice. The crowd at Freedom Stadium did not pay much attention to the presidential discourse, excited as they were by the appearance of the Pop Band 'Two Minds', just before the entrance of the athletes. But PKD, on the contrary, had carefully read the text that he had found in the archives of the Olympic Museum in Lausanne. Here is the passage that he quoted to our human fellows of the 22nd century:

> These New York Olympics, for which I have just solemnly declared the grand opening, announce a new era for sport. Even if all the members of the Olympic Committee are not totally convinced that the liberalisation of methods for performance enhancement in sport is in total conformity with the ethics and the spirit of sport, a majority estimated that the use of these substances was inevitable and therefore had to be ratified. For a very long time, my predecessors have tried to eradicate what was considered as a true plague in sport. The Committee now thinks it is time to accept the fact that these efforts have remained unfruitful. It is the conditions in the field that are forcing us to change our policies today. At my request, a specific commission was created to analyse the pertinence of an eventual liberalisation of methods for performance enhancement in sport. Doping is in no way a simple matter. In a sense, all the athletes can be considered doped, because every single one of them is an artificial being, built by artefact, equipped and helped by the existence of various devices. Why authorise a training session in altitude that will lead to an increase of erythropoietin while banning the direct injection of EPO? Why accept strenuous training that will provoke

muscle-fibre destruction and its regeneration through the natural creation of young myoblasts, while prohibiting the direct injection of DNA or stem cells into muscles, techniques that are providing similar results by enhancing muscular power? All sportsmen are doped, but some are trespassers because they are breaking the rules and laws of sport. The barometer can be set really high or really low, so that the cursor of prohibitions will fluctuate between very distant positions: the ancient and extremely rigid official position of our International Olympic Committee, forbidding hundreds of products without any practical possibility of enforcing bans, and the total laxness that characterised some of the national sporting federations. Doping is nothing but the logical consequence of the biomedical translation of the need to perform. The philosophy of sport that has, to this day, inspired the IOC's and WADA's repressive policies must be evaluated according to its consequences on the practice of sport and on athletes' well-being. We are forced to admit that this approach is an iniquitous source of injustice and that it threatens the athletes' health, forced to gulp down products acquired clandestinely and administered most of the time by staff with little medical competency. It is thus the will to minimise the risks for athletes' health and to introduce more justice in attributing sporting awards that the Committee members have decided to liberalise, for an eight-year trial period, the use of performance enhancing technologies under medical supervision. This role will be conferred to the AEP (Association for the Enhancement of Performance) – replacing WADA – to make sure the new regulation is being followed. *End of quote.*

The doping issue is at the very heart of the complex interrelationship of medicine and contemporary sport. It raises, in a very concrete manner, several fundamental questions. After quoting Perez's inaugural speech, PKD reviews the principal arguments in favour of liberalisation that had fuelled debates on this issue during the first quarter of the 21st century. During the first 20 years of this era, two main policies had been applied concurrently to limit doping: the outright prohibition of doping, and education and change in the social values associated with competitive sport. These two policies were a total failure. PKD calls to mind the arguments that had been advanced by partisans of the legalisation of doping during the first two decades of the 21st century, arguments that were present in Perez's speech.

1. The first one was pragmatically related to the inefficacy of controls and the persistence of doping despite the ban.

2. The second was linked to the ambiguous dual character of regulation: the official rules assimilated doping to fraud, whereas the unofficial rules, in some disciplines, pressured the athletes to engage in dubious conduct.
3. The last argument was directly related to the health of athletes. Perez's hope was that legalisation would more adequately protect the athletes' health by ensuring better medical supervision and by dragging the products out of the shadows.

Perez concluded his speech, using the arguments of the famous Belgian philosopher Bergil Toiho for whom a philosophy turned towards the respect of limits is not intrinsically more justified than one that glorifies the disregard for limits. The prohibition of doping was bringing a structural contradiction into competitive sport. The sportsman was being asked to surpass himself, while being prevented from doing so, on very controversial grounds.

The Olympics of the Eight

PKD illustrates the evolution of sport after the prohibition era with a description of the Shanghai 2064 Olympic Games. After Tokyo in 1960, Seoul in 1988, Beijing in 2008, Tokyo in 2020, and New Delhi in 2048, the Games were back in Asia. The IOC members had chosen Shanghai as the organising city for the 43rd Olympiad, and the presence in the city of Shanghai Transgenium's head office was not a total coincidence in their choice. When the Chinese star athlete Xiong Ni, gold medallist in the previous two marathons, took the Olympic Oath in the stadium freshly built by Shanghai Transgenium, only one word had changed compared to the text pronounced for the first time by the Belgian fencer Victor Boin at the 1920 Antwerp Olympic Games:

> We swear that we come to the Olympic Games as loyal competitors, respecting and abiding by the rules that govern them, and willing to take part in a spirit of chivalry for the honour of our countries and the glory of sport.

This was the original 1920 version. Only the end of the Oath was slightly modified in 2064: 'for the honour of the Eight and the glory of sport', swore Xiong Ni. The Eight represented the eight greatest biopharmaceutical firms whose athletes were ready to face each other in Shanghai where, for the first time, the competition between nations had been replaced by a battle between private companies. Some sports analysts at the time compared this development to the one that had marked the

Tour de France a century before, when national and regional teams had been replaced by brand promoting teams. Remember that Eddy Merckx, our Belgian national hero, donned the jerseys of Faema, Molteni and Fiat to ride the Tour, but never did he wear the national colours. The changes made in the organisation of the 2064 Games were, however, much more radical and came about as a consequence of a paradigm change in the philosophy of sport. The change in rules was therefore not contested, except by a minority of intellectuals and a handful of spectators who were feeling nostalgic for a romantic vision of sport that had probably never existed.

In this new paradigm, biological modifications had become so efficient that they held a promise of insane performances increasing the scope of what humans had previously been able to achieve. For athletes though, pain remained the privilege of daily training sessions, led by former champions. But the idolised athlete was now entrusting the Firm with the task of gradually improving the physiology of his organism. The dream of every athlete was to belong to the elite of sport by being signed up to one of the Eight. The physician–trainers of the Eight had mastered the art of adapting the athlete's multiple biological variables to competition. Scientists and engineers were constantly providing new weapons to boost the performance of the athlete's body. Research and technoscientific development were being integrated, more than ever before, to the practice of sport. The body had become similar to the engine of a Ferrari that engineers adjust with mechanical precision on the eve of a Grand Prix. Except that engineering the living is a far more complex and delicate art than that of the mechanical engineer.

Some fans, disgusted by this Promethean development, had turned away from professional competitive sport. They were wondering what had become of the sportsman, reduced to the role of a passive element at the heart of a system that conditioned and totally surpassed him. According to them, the possibility of manipulating the organism and simultaneously transforming athletes into winning machines threatened to desecrate the human being. It was a matter of urgency to take these upheavals into account and define new objectives, even if this meant the suppression of professional sport competition. A majority though remained faithful to the new formula. As for the young generations who had never known any other form of competition, they were simply coping with the fact that 'the athletes were merely the actors of a show', too fascinated by the struggle between these modified bodies to reject it on past values. The representation of sporting activity had oscillated between two distinct philosophies perfectly described by the

Belgian thinker Bergil Toiho, two anthropologies or conceptions of Man: the way of balance that strictly respects the limits imposed by nature and the way of transcendence that aspires to surpass these same limits. Little by little, the philosophy of artifice oriented towards the surpassing of limits took over the society in which the naturalistic philosophy of respect had once shone so brightly. The latter had, indeed, completely lost touch with the spirit of the first half of the 21st century techno-liberal society.

Debate concerning the evolution of sport had long been nourished by the more or less implicit valuation of the natural against the artificial, and by the positive evaluation of some techniques as opposed to others.

The arrival of Antonio Perez at the IOC in 2021 gave a new spirit to the philosophy of sport. The sports federations – encouraged by the athletes themselves and the majority of their audience members – understood that it was important not to condemn the inventive genius of man in blind defence of a naturalistic philosophy of sport. The old opposition between nature and artifice, valued respectively as good and bad, was not only open to criticism because of the hierarchy it aimed to impose, but also because of its intrinsic pertinence. As Bergil Toiho had written, 'is there anything that is not, to a certain extent, artificial, namely techno-cultural, in our societies?' What in the human world has remained perfectly natural? Most of the techniques across all different aspects of life had been designed to enhance performance. The problem was thus that of distinguishing between the good, legitimate technologies and the others. For a long time, only external technologies had been considered acceptable. External technologies are those related to sporting equipment, those that enhance performance without modifying the athlete's body: better shoes, better surfaces, stadiums with climate control systems, starting blocks (that, in the 1930s, had replaced the holes dug by the athletes themselves with a shovel in the track), better bikes, appearance of derailleur gears, toe clips, automatic shoe fixings, lightweight carbon-rigid bikes, aerodynamically designed models, lenticular wheels, and so on. These new technologies had always been accepted, except when their transformation of the nature of the game was deemed too radical. PKD gives the counter-example of the double rope rackets used by Jimmy Connors at Wimbledon, before they were banned by the International Tennis Federation.

The technologies that directly affected the human body were first considered illegitimate. We have seen that in 2024 the use of pharmacological products was tolerated for pragmatic reasons. The fears were directed at that time towards the deep and irreversible modifications of the human body. When this worry was at its peak, it was expressed by speculation concerning a future eugenics for sportsmen. It was estimated, with good reason, that if positive eugenics was to surface in the course of the 21st century, there was some probability that this would happen in the field of competitive sport. The relatively easy targets of that activity offered a predestined ground for the first experiments in genetic enhancement. Unfortunately, PKD in his inaugural conference is not very explicit about the physiological data in relation to the modification of athletes. Indeed, the rare technical terms to which he refers come from the specialised scientific vocabulary of the 22nd century and are therefore of very little help to us. What can clearly be claimed is that competition in 2144 had become largely techno-scientific, a contest between laboratories implying radical modification of the athlete's body.

Epilogue

As he left the Raymond Goethals Stadium after the men's 100 m final, PKD had goose bumps. He was preparing his speech for the bicentenary celebration of the creation of the University of Brussels' Institute for the Philosophy of Sport, and the work he had done to document his research had shaken his certainties. What conclusions should we draw from these transformations in sport? Did we not underestimate the consequences that might arise from the liberalisation of performance enhancing techniques? Did Antonio Perez not commit a historical mistake when he legalised doping? PKD was particularly concerned about the future of his son who was playing in the Brussels' Junior Rollerball team. It was precisely at this stage that the coaches were selecting the best players to form the elite teams. A handful could even hope to play one day in the adults' Premier League team. But was this an enviable fate for his son? Furthermore, if Tom were selected to join the team, how could he refuse? To his great fortune or misfortune, his son was quite gifted according to the evaluation grid, so the coach had asked to meet PKD. What did the coach want? The idea that enhancing technologies might be applied to his son disturbed PKD. Sport could probably be considered the laboratory for those changes that will ultimately contaminate other sectors of society. What will happen to the human being, taken as a whole, if we start to enhance him?

While thinking about the fears he was experiencing for the future of his own son, PKD realised that his idealistic theoretical speech, valid without a doubt in the ethereal world of reason, was far less evident in the realm of feelings. His reason could find no strong argument to criticise either Antonio Perez's decisions or the ethical and philosophical justifications of Bergil Toiho. But, in his inner self, he was silently hoping that Tom would not be selected to pursue the road with the best. He did not want to see Tom enlisted in the performance enhancement programme that the Brussels' Rollerball club imposed on its athletes from their teenage years. And yet PKD truly appreciated the entertainment that competitive sport offered. Moreover, he could not prevent himself from thinking that authentic sport was what he practiced quite clumsily, every Saturday afternoon in the football stadiums of Brussels' southern suburb. PKD did not dream of his son joining the sporting elite of the Eight. He preferred that Tom remain a Saturday amateur player, just like himself.

He had also been quite frightened to learn that some of his colleagues had chosen to modify their child's genome at the embryonic stage by adding the new artificial chromosome commercialised by Shanghai Transgenium: a complex DNA providing reinforced immunity, an anti-oxidation system against the premature aging of cells, and finally a protection against certain forms of cancer. Nothing radically terrible, but this was just a starter. He was happy to have been born sufficiently early in history to avoid having to make that kind of decision for Tom. He also had the feeling that he had talked too much, whereas he was usually not very talkative. His right arm had stiffened from having maintained his elbow on the table for such a long time. As for his left hand inserted in the pocket of his pants, it held the two precious tickets that he had been able to obtain for the Galactic League match of 17 October opposing Milan AC to the Brussels' Daring of Jeff Koens. Tom would be enchanted. Just like his father, he loved sport.

Notes

This chapter draws from ideas developed in French in two previous publications: Missa, J.-N., 'Sport, enhancement and the inefficacy of anti-doping policy', *Bioethica Forum*, 2011, 4, No. 1, 14–6; Missa, J.-N., 'Dopage, médecine d'amélioration et avenir du sport', in Missa, J.-N. et Nouvel, P., *Philosophie du dopage* [Philosophy of Doping], Paris 2011, pp. 35–84.

1. See the full transcript of Lance Armstrong's interview with Oprah Winfrey: http://armchairspectator.wordpress.com/2013/01/23/full-transcript-lance-armstrong-on-oprah/. Accessed January 26, 2013.
2. See the *Statement from USADA CEO Travis T. Tygart Regarding the U.S. Postal Service Pro Cycling Team Doping Conspiracy* (http://cyclinginvestigation.usada.org/).

References

—— (2001) 'Olympics– Samaranch: Doping is a Fact', *New York Times*, 2 July 2001, http://www.nytimes.com/2001/07/02/sports/olympics-samaranch-doping-is-a-fact.html. Accessed September 17, 2013.

Belton D. (2001) 'Ben Johnson', in *Reputations*, documentary series (UK: BBC, 1994–2002).

Chambers D. (2009) *Race against Me. My Story* (London: Libros International).

Chapatte R. and Gavinet L. (1962) 'Attention danger', in Goddet J. and Marcillac R. (prod.) *Les coulisses de l'exploit*, documentary series (France: RTF and Pathé cinéma, 1961–1972).

Friedmann T. (2002) 'Potential for genetic enhancement in sports', transcript, 11 July 2002, in *The President's Council on Bioethics*, www. bioethics.gov.

Friedmann T, Rabin O., Frankel MS (2010) 'Ethics. Gene doping and sport', *Science*, 327, 647–8.

Kayser B., Mauron A., Miah A. (2005) 'Legalisation of performance-enhancing drugs', *Lancet*, 366, S21.

Kayser B., Mauron A., Miah A. (2007) 'Current anti-doping policy: a critical appraisal', *BMC Medical Ethics*, 8(2). DOI:10.1186/1472-6939-8-2, published online March 29, 2007.

McPherron A.C., Lawler A.M., Lee S.-J. (1997) 'Regulation of skeletal muscle mass in mice by a new TGF-ß superfamily member', *Nature*, 387, 83–90.

Miah A. (2005) 'Gene doping: a reality, but not a threat', 2 May 2005, www.andymiah.net.

Nadal, R., BBC Sport, 12 February 2009, http://news.bbc.co.uk/sport2/hi/tennis/7885314.stm. Accessed September 17, 2013.

Savulescu J. and Foddy B. (2007) 'Ethics of performance enhancement in sport: drugs and gene doping', in Ashcroft R.E., Dawson A., Draper H. and McMillan J.R. (eds) *Principles of Health Care Ethics*, (2nd edn) (London: John Wiley & Sons), pp. 511–20.

Savulescu J. and Foddy B. (2009) 'Le Tour and Failure of Zero Tolerance: Time to Relax Doping Controls', in ter Meulen R., Kahane G. and Savulescu J. (eds) *Enhancing Human Capacities* (Oxford: Wiley Blackwell).

Wenner M. (2008) 'How to be popular during the Olympics: be H. Lee Sweeney, gene doping expert', *Scientific American*, 15 August 2008, www. scientificamerican.com.

12
Adaptation and Emortality: Human Enhancement in *Tales of the Biotech Revolution*

Brian Stableford

Before 1980, futuristic fiction made little use of speculative biotechnologies as a potential means of human enhancement, and the most famous early-20th century novel anticipating the widespread future application of biotechnologies, Aldous Huxley's *Brave New World* (Huxley, 1932), was a vitriolic satire that viewed the prospect of technological interference with human nature as implicitly horrific. Although most of the speculative motifs featured in that text were appropriated from an essay whose tone was resolutely optimistic, J.B.S. Haldane's *Daedalus; or, Science and the Future* (Haldane, 1923), it is notable that other direct fictional extrapolations of the same essay were anxiously negative in tone. These included those by Haldane's friend Julian Huxley (Aldous Huxley's brother), 'The tissue-culture king' (Huxley, 1926), and Haldane's sister, Naomi Mitchison, *Solution Three* (Mitchison, 1975) and *Not by Bread Alone* (Mitchison, 1983). Although use of biotechnological themes in futuristic fiction increased markedly after 1980, the general tone of such fiction remained hostile, often hysterically so.

J.B.S. Haldane had anticipated this negative reaction, and had taken the opportunity in the introduction to his own essay to argue that 'biological inventions' are invariably seen, at their inception, as innately hideous perversions of 'nature', although some of them eventually come to seem entirely natural once they become familiar. What opprobrium, he asks, must have been heaped upon the first person to advocate milking cows? He was well aware that innovations associated with biology – especially when they have some relevance to the digestive system, the excretory system or sexual biology – often generate a quasi-reflexive reaction commonly known as the 'yuck factor'. He contented himself with pointing out that such factors usually lose their force quite rapidly, but might well have pointed out that everything we now think of as typically human can be conceived and explained as extrapolations of two primitive biotechnologies: cooking and clothing.

One of the key biotechnologies featured in the anticipatory section of Haldane's essay was what he called 'ectogenesis': the possible future development of artificial wombs in which human egg-cells could be implanted, allowing developing embryos to be much more carefully monitored and perhaps subject to modification. This became the central image of Aldous Huxley's satire, which begins with a description of a 'hatchery' and employs embryological manipulation to produce a rigidly stratified society of alphas, betas, gammas, deltas and epsilons, the members of each caste being designed to fulfil various sets of occupational functions. We have now had the opportunity to see the first phase of this development within our own lifetimes, in the development of techniques of *in vitro* fertilisation and selection – which have, as Haldane anticipated, rapidly progressed from arousing reflexive horror to near-universal acceptance.

The broader reflection of Haldane's speculations in popular fiction provides a graphic illustration of the fact that writers of fiction invariably find more dramatic potential in the initial horrific reaction than in the eventual meek acceptance. In fiction, in all strata of the literary marketplace, biotechnology has been given exceedingly bad publicity, but this is at least partly due to the fact that all 'intrusive fantasies' – the broader category that includes accounts of technological innovation – have a 'natural story arc' in which the banishment of the intrusion/innovation restores an ordered *status quo*, thus providing a satisfactory sense of narrative closure. The convenience and popularity of this story arc, combined with a natural temptation on the part of authors pandering to their audience to exploit the yuck factor, goes a long way to explaining the pattern of usage of biotechnological innovations in futuristic fiction.

In all fairness, however, it must be admitted that *all* kinds of potential human enhancement have had a great deal of bad publicity in fiction, again in all strata of the literary marketplace. Even the most superficial analysis of the literary history of 'superhumanity' reveals an extensive category of cautionary tales, even with reference to enhancements that might seem to be purely beneficial, including the possibility of extending the human lifespan – one amendment that might be expected to have a good deal of support, given the apparent dread and reluctance most people exhibit when it comes to dying.

If, as existentialist philosophers suggest, *angst* arising from the consciousness of inevitable death is a cancer corroding the very core of human existence, why is it that the literary images of hypothetical individuals who need not die routinely depict them as tragic figures whose longevity is an unmitigated curse? Perhaps it is a corollary of Aesop's fable of the fox and the grapes, in which the fox consoles himself with the notion that the fruits which remain out of his reach are probably sour. Since we must, in fact, die, and cannot like the idea, perhaps it is in our interest to pretend that the prospect of an extended life, even in comfortable circumstances, would be even worse than the experience of a life that is brutish and short?

The benefits of modern hygiene and medicine have already added a further decade to the traditionally recognised human lifespan of three-score years and ten in developed countries, and have drastically reduced the number of people who die prematurely before reaching that limit. We do not know, as yet, how far future developments in biotechnology might extend the limit, or how rapidly, but it would be foolish to deny or ignore the possibility. Logically, the more that science allows us to understand the various processes of aging, the better our chances will become of building efficient defences against them, thus obtaining greater longevity. Perhaps, eventually, that longevity might become potentially unlimited – a condition that Alvin Silverstein has labelled 'emortality' in preference to the more commonplace 'immortality,' because a person thus gifted would still be subject to the possibility of violent destruction, and would only exist so long as fatal accidents could be avoided (Silverstein, 1979).

The possible advent of emortality, accompanied by other hypothetical consequences of sophisticated biotechnology, has long been a preoccupation of mine, expressed in such futurology books as *Future Man* (Stableford, 1984) and *The Third Millennium* (Stableford and Langford, 1985), and in the core of my endeavours as a science fiction writer – a core that currently amounts to eleven novels and six collections of short stories. Having long been an admirer of Haldane's essay, and having taken to heart the argument contained in its introduction, I have always considered it as a moral duty not to yield to the conventional pressure to employ speculative biotechnology as a means of manufacturing monsters – and, in particular, not to represent emortality as something essentially undesirable. I have always attempted to oppose the yuck factor, not only by not pandering to it in my plotting, but by consistently attempting to produce images of the future in which various kinds of biotechnologies are seen as entirely normal by the characters in the stories: aspects of everyday life as automatically taken for granted as cooking food, wearing clothes and all the other blatantly 'unnatural' things people nowadays do.

In stories dealing with biotechnologies that are currently under development, such as 'What Can Chloë Want?' (Stableford, 2004) – which features a slightly modified version of xenotransplantation – I try to be extremely scrupulous in emphasising the positive aspects of such developments. While I have never sought to conceal the inevitability that any new technology might give rise to unfortunate consequences as well as fortunate ones, I have always tried to give due emphasis to the latter, in spite of the narrative difficulties involved – there is, of course, far more dramatic potential in the description of things going wrong than in the representation of things going well.

I have occasionally used fictional frameworks to explore the possibility of awkward forms of emortality. In 'And He Not Busy Being Born...' (Stableford, 1991) for instance, I describe a society in which the price of extreme longevity is that the body's development must be arrested before puberty, so that the hypothetical emortal individuals resemble our nine year-old children in physical terms. In 'The Age of Innocence' (Stableford, 2004), I describe a society in which a lifespan much longer than seventy years is achievable, but the latter phases of it involve a gradual mental decline and physical diminution, so that the 'second childhood' of senility lasts for several decades, and the burden of caring for the innocents in question routinely falls, albeit temporarily, upon children. In *Zombies Don't Cry* (Stableford, 2011c), emortality might be an attribute of those resurrected from the dead by means of a somatic transformation, although the possibility remains untested, whereas it still remains out of reach – thus far – of the living. A more extreme problem of a broadly similar kind provides the crucial revelation in the climax of *The Cassandra Complex* (Stableford, 2001), in which the first method of prolonging life to be discovered carries a penalty so nasty that the method has to be shelved until a better one comes along – an outcome represented as a tragedy rather than some kind of divine or quasi-poetic justice. The main thrust of my fictional explorations of the possibility of biotechnological longevity has, however, been speculation regarding the adaptations of social organisation that would be forced upon a world in which death is rare and the control of reproductive rates a necessity.

Most of my stories of this broad type take it for granted that most children in a society of emortals would be raised by collective households involving more than two adults, with no necessary biological relationship between children and parents. The most detailed descriptions of such hypothetical households are contained in 'The Invisible Worm' (Stableford, 2004), which depicts some of the potential petty tribulations of that kind of parenthood, and *The Dragon Man* (Stableford, 2009c), which attempts to examine the particular problems of adolescence as they might be manifest in such a society. My stories in this vein usually retain the assumption that the right to found a family recognised in the UN charter will still be maintained, but assume that it will usually be exercised posthumously, by means of sperm and eggs placed in long-term cold storage. Alternative scenarios, including the one developed in 'The Pipes of Pan' (Stableford, 2004), in which the cost of maintaining nuclear families in a world without death is that the ages of children must be artificially maintained at a particular developmental stage indefinitely, are represented as precarious as well as rather ridiculous.

I confess that I have not devoted a great deal of literary attention to the problem most frequently anticipated by most negative accounts of the prospect of emortality: that of unbearable tedium. Karel Čapek's *The Makropoulos Secret* (Čapek, 1925), for instance, features a woman only a few hundred years old, who feels that she has seen and done everything possible so many times that all experience has lost its savour and its meaning. The reason I have not made much effort to refute this argument is that it seems to me to be manifestly absurd. Even if the world and the consequent scope of human experience were not changing as rapidly as they are, human existence surely offers sufficient abundance of experience to sustain most individuals through millennia rather than mere centuries, and, even in the unlikely event that Earthly society were some day to attain a Utopia of universal contentment, in which no one saw any need for any further progress, there is no reason to take it for granted that the society in question would be an inescapable prison.

My own opinion is that the opposite is more likely to be the case, and that human experience, especially the human experience of other human beings – whose variety is immense if not illimitable – is potentially so complex and subtle that tedium is only possible in common mortals, as a consequence of intellectual cowardice and despair engendered by *angst*. It only requires a courageous change of attitude and a healthy measure of inquisitiveness to see potential emortality as a wonderland of infinite opportunity, and I cannot imagine that many people who had that fruit potentially within their reach would assume it to be sour. Any that did make that assumption would, of course, remain free not to grasp and taste it – emortality carries no compulsion to continue living, if its beneficiaries would rather not do so. The fact that so few mortals commit suicide is perhaps cause for surprise, given the dire circumstances in which most of them live, but it certainly suggests that suicide would not be any more popular as an option in a world in which people were capable of longer life.

There are, however, other problems frequently identified in hypothetical fiction about longevity that seem to me to require more serious attention. One is the problem of inheritance. If the old need not die, then they need not pass on their roles, or their possessions, to the young, who would have to wait a great deal longer to take over the reins of society and its accumulated wealth. Generational tensions and conflicts might, therefore, be expected to intensify, perhaps preserving motives for violence and murder even in societies in which everyone's life is comfortable. Although that is a useful source of melodramatic capital – which I have exploited in such tales of futuristic murder and mayhem as *Inherit the Earth* (Stableford, 1998) and *Les Fleurs du Mal* (Stableford, 2010a) – I have been reluctant to take too much advantage of it, because it runs contrary to my ideological mission. In most of my works I make the assumption that the young would eventually learn to be patient in matters of inheritance, that being a valuable lesson that they could carry forward into their prime and beyond, along with the corollary assumption that the young would continue to maintain the search for new existential opportunities and new horizons, while allowing the old to keep their existing roles and possessions. My most detailed fictional account of future history, *The Fountains of Youth* (Stableford, 2000), takes this pattern for granted.

At present, of course, we live in a world dominated by short-term thinking, in which such mottoes as *carpe diem* and 'live fast, die young, leave a beautiful corpse' have a certain sad appeal. The world is mostly run by the old, who appear to be hell bent on spending their inheritance as rapidly as they can, leaving nothing to future generations but debt and debris. It is conceivable that this would not change even if people could look forward to potential lifespans measured in hundreds of years rather than decades, but the logic of natural selection suggests that those individuals capable and willing to adjust their attitudes to longer-term goals would eventually inherit the world, even if they had to fight a long battle to get the will through probate. All my works of futuristic fiction assume that things will get worse before they get better, and that Utopian ambitions will only achieve a fair hearing after the world's economy and ecology have suffered a fatal Crash (when I began writing, that still qualified as an anticipation, but now the Crash in question is so well-advanced that only idiots and people in pathological denial – who, alas, still constitute a majority – can refuse to see it).

Another problem that requires some hypothetical consideration in the serious contemplation of the possibility of technologically enhanced longevity is that of memory. To what extent, we are obliged to wonder as we contemplate the prospect of emortality, would long-lived human beings recall their increasingly distant pasts? Would they be forced to become adepts in the art of forgetfulness, retaining only the vaguest impressions of spans of time they have left centuries behind, or would they cultivate more powerful and capacious memories than the ones we have, perhaps with the aid of artificial augmentations of some kind? Most of my representations try to have it both ways, assuming that certain key events would continue to be remembered, even over vast chronological distances, while others would fade and be compacted, but that people would increasingly have recourse to various kinds of memory aids, in order that they could continue to construct long and complex biographies, such as the one that forms the main narrative of *The Fountains of Youth* (Stableford, 2000) and is further augmented in *The Omega Expedition* (Stableford, 2002b), extending over several hundred years. The problem of memory is also a significant secondary theme in *Les Fleurs du Mal* (Stableford, 2010a).

The question of how human individuals and societies might employ longevity or emortality is, of course, largely dependent on what other kinds of opportunities might be opened up to them by sophisticated biotechnologies. Indeed, literary images produced after 1980 have frequently made emortality part and parcel of more complex images of 'posthumanity', in which the long-lived individuals may be considerably modified anatomically and physiologically, either by genetic engineering, cyborgisation or complex combinations of the two. When I was commissioned to write my first futurology book on this subject, *Future Man* (Stableford, 1984), in the early 1980s, I was asked specifically to give elaborate consideration to the potential adaptive radiation of the human species, in response to opportunities of anatomical and physiological modification.

Previous imaginative fiction had devoted considerable attention to the possibilities of adapting humans for life in extreme or alien environments: for an amphibious life based under water, for example, or the equipment of humans with wings to facilitate transportation. The most notable pioneering work of science fiction extrapolating this line of thought, a story-series written in the 1950s by James Blish and collected in *The Seedling Stars* (Blish, 1956), had proposed a principle of 'pantropy', by means of which genetic engineering might allow human beings to colonise a vast range of environments distributed throughout the galaxy.

Although I began my writing career in an era when the myth of the Space Age was not yet devoid of its last vestiges of rational plausibility, I considered that myth to be essentially obsolete by the time I began to compose *Tales of the Biotech Revolution* in earnest. My stories in that vein take it for granted that human expansion beyond the Earth's surface is bound to be an extremely slow and difficult process, involving the painstaking exportation and reproduction of the environment to which we are already adapted rather than the discovery of new ecospheres to which we might adapt ourselves technologically.

For this reason, my work generally takes it for granted that the one significant gross anatomical modification that our imminent descendants are likely to make will consist of physical and physiological adaptation to life in low-gravity and zero-gravity environments, gravitational attraction being one thing that we cannot easily export in moving away from the Earth's surface. The modified individuals in question, called 'fabers' in my stories, effectively have four arms instead of two arms and two legs; they are most extensively featured in *The Fountains of Youth* (Stableford, 2000). Most of the physiological modifications envisaged in my images of future society are much subtler in anatomical terms.

Any contemplation of the present situation in developed countries with regard to surgical enhancement is bound to convince the observer that much of the near-future demand for applied biotechnologies will be in the field of cosmetic enhancement. I have attempted to bring possibilities of that sort into sharp focus in a number of stories, most notably 'Cinderella's Sisters' (Stableford, 1991), in which two 'ugly' sisters compete in using new biotechnologies in order to enter into a fierce beauty competition, not merely with one another but with a hypothetical ideal, and 'Skin Deep' (Stableford, 2007b), in which the protagonist's desire to keep up with ever-changing fashions in female beauty is eventually confounded by an exotic crime of passion. Cosmetic enhancement is a significant secondary theme in many other stories that take such advancements for granted, most notably *Les Fleurs du Mal* (Stableford, 2010a), which features a protagonist named Oscar Wilde and makes extensive ironic reference to *The Picture of Dorian Gray* (Wilde, 1891). In accordance with the logic of the situation, I have assumed that such technologies will be used, at least in the short term, to bring people into line with pre-existent standards of beauty rather than to explore more exotic variations, but that once everyone can meet pre-existent standards of beauty, the quest for distinctiveness is bound to engender bolder experimentation.

I have also tended to be modest in extrapolating technological augmentations of the human sensorium, although I have been prepared to grant that when the day arrives that artificial eyes and ears can perform better than natural ones, there will be a seductive appeal in the possibility of making the trade. I cannot help suspecting, however, that if there were no costs in expanding the range of electromagnetic radiations perceived by the eye, or the range of sounds perceptible by the ear, then natural selection would not have restricted them, and that it might be the case that the spectrum of sensory perceptions with which we are already equipped may be not far removed from the optimum with which our brains can presently cope.

In time, we might be able to improve our brains in that respect, but in the shorter term, it has always seemed to me that the human organ that might gain most from technological augmentation, and with respect to which such augmentation might turn out to be relatively easy, is the epidermis. In such stories as 'The Skin Trade' (Stableford, 2007b), I have focused explicitly on the possibility of developing highly sophisticated garments, perhaps only a few molecules thick, which could simply be overlaid on the body's existing outer layer – including the lining of the gut – in order to enhance its protective capabilities in various different ways. Many of my stories set in more remote futures take it for granted that such 'surskins' are a familiar aspect of everyday existence; their decorative potential is given specific attention in *The Dragon Man* (Stableford, 2009c), but they are more generally featured as invisible assistants.

In much the same way that the human skin might be augmented unobtrusively, I have tended to assume that most other augmentation of the body's powers of maintenance and self-repair would be virtually invisible, mostly taking the form of 'suites' of 'internal nanotechnology'. Such 'suites' might, at least in the short term, be an invaluable aid in procuring and sustaining longevity, perhaps with dramatic side-effects such as those featured in 'Busy Dying' (Stableford, 2007a), whose protagonist conceives a determination to test the life-saving potential of internal nanotechnology to the limit. I have routinely assumed that such nanotech 'suites' would be subject to continual replacement by improved versions, thus providing the basis of an 'escalator effect' by means of which beneficiaries of an initial technology of longevity that adds, say, twenty years to the average lifespan, might thus be enabled to benefit from the next generation of such technologies, which might add a further fifty years, and so on, leading them incrementally to full emortality. The potential effects of an *angst*-opposing consciousness of that possibility are given substantial passing consideration in *Inherit the Earth* (Stableford, 1998).

In general, I have steered clear of the more extreme metamorphoses that future biotechnology might permit, for want of the imagination required to envisage potential motivations for such transformations, although many of my future scenarios – including *Les Fleurs du Mal* (Stableford, 2010a) – carry forward from *The Third Millennium* (Stableford and Langford, 1985) the idea that small islands might be taken over by artistically inclined 'creationists' intent on devising and constructing exotic ecospheres. The individuals subjected to metamorphosis in such stories are mostly animals rather than people, however, including the narrator of 'Dr. Prospero and the Snake Lady' (Stableford, 2009a), an enhanced orang-utan jealous of her 'father's' love and attention. I have made one significant exception in respect of the possibility that radical transfiguration might offer a palatable alternative to otherwise unavoidable death. In 'The Great Chain of Being' (Stableford, 2009d), I describe an admittedly bizarre scenario in which brain-death remains unavoidable, but the continuation of the flesh can be achieved by metamorphosis into a tree, with the result that 'human trees' begin gradually to displace other species and to take over the Gaian responsibility of maintaining and managing the ecospheric environment. Similar but more limited hypotheses are developed in 'The Tree of Life' (Stableford, 2007b), which features a gene-appropriating 'Haeckel tree'; 'Another Branch of the Family Tree' (Stableford, 2004), in which a geneticist attempts to preserve her dead twin sister by incorporating her genome into that of a tree, but then must fight to save the tree from the woodcutter's axe; and *The Undead* (Stableford, 2010b), in which organic tombstones preserving memorial records of deceased individuals begin campaigning for more active bodies and quasi-human rights.

In general, I have assumed that the most dramatic biological transformations that our imminent descendants are likely to carry out will not be human modifications at all, but modifications of plants and animals, carrying forward the long-standing tradition of biological manipulation of crops and livestock. The history of biotechnology demonstrates that plants are easier to modify technologically than animals, and I have assumed that this principle will continue to hold, although I have made the corollary assumption that there will probably come a day when artificial processes of energy-fixation will be developed that are even more effective than the natural technology of photosynthesis, thus allowing the burden of primary biological production to be taken over by industry.

The possibility that such industrial processes might provide invaluable new niches for parasitic organisms that natural selection would inevitably produce in response, is featured in 'Hidden Agendas' (Stableford, 2007b), which features a controversial heroic attempt to help insect species survive a particularly drastic ecological Crash, further extrapolated in a subplot of that story's direct sequel, *Xeno's Paradox* (Stableford, 2011b). In the shorter term, however, the genetic modification of crop plants offers abundant scope for speculation, if one is prepared to grant that Haldane's principle of acceptance will soon triumph over the current hostility to such modifications manifest in developed countries. Accounts of near-future plant modification provide the primary subject matter of several of my *Tales of the Biotech Revolution*, including 'Rogue Terminator' (Stableford, 2007b), which features the kind of 'terminator technology' briefly made notorious by the ill-fated Monsanto; 'The Home Front' (Stableford, 2007b), which features a stock market bubble in genetically engineered potatoes similar to the great Dutch tulip boom; 'Following the Pharmers' (Stableford, 2009d), in which a scientist tries to equip flowering plants with the independently motile seeds that they never contrived to develop unassisted; and 'Some Like It Hot' (Stableford, 2009b), which explores certain possibilities of genetically engineered hemp and kelp in the context of global warming.

With respect to animal modification I have assumed that it will be easier to modify invertebrates than vertebrates, and have put invertebrate modification in the foreground of a number of stories, including 'The Invertebrate Man' (Stableford, 1991), which adds a new twist to the cliché of giant spiders; 'The Growth of the House of Usher' (Stableford, 1991) – subsequently expanded and elaborated as *Nature's Shift* (Stableford, 2011a) – which suggests possible artificial modifications of the metamorphic potential of insect larvae; and 'The Road to Hell' (Stableford, 2007a), which extrapolates the technological potential of genetically modified hookworms. On average, these stories allow anxiety to show through more frequently and more graphically than stories of plant modification – a pattern further extrapolated in accounts of unintended side-effects of attempted human modifications, such as 'The House of Mourning' (Stableford, 2004), about genetic modification for sexual purposes; 'The Milk of Human Kindness' (Stableford, 2004), in which the choices to be made from a wide range of genetically enhanced milk cause marital discord; 'The Last Supper' (Stableford, 2004), which features a restaurant specialising in genetically modified food; and 'The Trial' (Stableford, 2009a), in which a new treatment for Alzheimer's disease turns out to work a little too well on one of the patients in the crucial clinical trial.

I have, however, conscientiously restrained myself from producing too many stories of this general type, for fear of contributing to the problem that my work in this vein is attempting to solve. In stories that feature one of the more familiar themes of animal enhancement – that of equipping animals with quasi-human intelligence, perhaps supplemented with human appearance – I have been particularly careful to emphasise the positive; key examples include 'Out of Touch' (Stableford, 2007b), which features a monkey genetically enhanced in order to serve as an assistant to the elderly; 'Snowball in Hell' (Stableford, 2004), in which a method of 'applied homeotics' works spectacularly well on pigs; and *Xeno's Paradox* (Stableford, 2011b), which extrapolates the potential of applied homeotics to a more exotic extreme.

No writer is immune to the seductions of melodrama, which are harmless in stories set in the past, since the reader already knows that the past was successful in producing the present, but surely ought to be considered toxic in accounts of the future, in which our hopes and fears are necessarily in conflict. If one takes individual stories one at a time, there is no particular harm in accounts of technological endeavour going awry, producing problems whose amelioration can provide a satisfactory sense of narrative closure. If one looks at a vast array of such stories as a whole, however, their commitment to convenient story arcs and easy narrative closure tends to give rise to the impression that all technological innovation is direly dangerous, and best avoided: a phenomenon that Isaac Asimov has labelled 'the Frankenstein complex'.

Thanks to Asimov, who explained in the introduction to *The Rest of the Robots* (Asimov, 1964) that he had written his long series of robot stories in determined opposition to the Frankenstein complex, I was aware before I started my biotech series, not only of the problem in question and the ideological necessity of coping with it, but also of the difficulties inherent in that policy. Any writer who deliberately shuns the easy narrative currency of conventional melodrama and the standard story arc of intrusive fantasy is inevitably obliged to find alternative forms of narrative energy and drive, and that is not an easy task. It is particularly difficult to satisfy the demands and expectations of readers of popular fiction while scrupulously avoiding the most precious melodramatic resource that has evolved in order to satisfy those demands and expectations.

The one advantage that writers of futuristic fiction possess in this regard is that exotic ideas are sometimes sufficiently striking to sustain stories merely by their display, without any conspicuous embellishment by plotting. If such ideas are to be properly explored, however, especially to the extent that they become familiar – as the most plausible ones do with an inevitable rapidity – they require more narrative muscle to sustain and strengthen them. My self-appointed duty not to overaccentuate the negative possibilities of the technologies whose potential I am exploring thus creates a corollary demand for the continual deployment of story arcs that are bound to seem unusual and unsatisfying to readers accustomed to the formulas of melodrama.

Story arcs featuring the creation and eventual neutralisation of threats – the staple fare of crime and thriller fiction – are not, of course, the only standardised formula to be found in popular fiction. There are also story arcs that feature achievement, in which characters are rewarded for persistence and ingenuity by the eventual acquisition of some crucial success, in terms of wealth, a sexual relationship, wisdom or all three. This is the staple fare of generic romantic fiction, and it might seem, on the surface, to offer more scope to writers intent on painting the future possibilities in a positive light. Not unnaturally, I have made abundant use of it in my *Tales of the Biotech Revolution*, many of which are accounts of hard-won individual triumph. There is, however, an important sense in which this source of narrative energy, too, is tainted when it is transplanted from past and present-day scenarios into the future. Future possibilities not only might, but almost certainly will, transform the goals that are presently regarded as the key potential achievements in human life. The contemplation of human history immediately reveals that our notions of wealth and satisfactory sexual relationships have changed dramatically over time – and that our notions of wisdom have altered even more drastically.

In his philosophical novel about the limits of anticipation, *Sur la pierre blanche* (France, 1905), Anatole France argued that literary attempts to predict the future were doubly doomed, firstly by the impossibility of anticipating the as-yet-invisible trends and as-yet-unmade discoveries that will motivate the transformation of present-day society, but also by the ironic fact that anyone who successfully anticipated the morality of the future would necessarily seem guilty of flagrant immorality by the standards presently in force, and would be condemned for it by readers and critics alike.

Many writers of futuristic fiction react to this problem warily, by carefully preserving key elements of contemporary moral prejudice in their future scenarios and, indeed, using the future as an arena into which to project their moral prejudices. It is, in fact, very difficult to avoid doing that, even for those writers who take some delight in offending current moral sensibilities they deem to be mistaken. If it is very difficult to anticipate the values of the future, however, it is frankly impossible to anticipate the wisdom of the future; the one thing of which we can be sure is that it will not be the wisdom of the present. Because of these difficulties, there has always been a certain unease underlying my *Tales of the Biotech Revolution*, which has made me as reluctant to use conventional improving endings as I am to use conventional normalising endings.

I do occasionally employ literal enrichment, sexual success and presumed enlightenment as narrative rewards, but I always try to do so in a sarcastic spirit that casts doubt on the permanent worth of the notions we presently have of wealth, moral rectitude and good thinking. Like many writers of *contes philosophiques*, however, I have more often been drawn to blackly humorous *conte cruel* endings, which deliberately violate reader expectations with regard to moral and intellectual endorsement, embracing a calculated and deliberate perversity in the interests of challenge. In so doing, I try not to lose sight of the fact that I really am trying to be cruel in order to be kind, and that my apparent cynicism is directed toward Utopian ends but that is always an awkward tightrope to walk, and I am always open to the suspicion that I am merely – as one reviewer once described one of my protagonists – a 'facetious churl'.

There would be little profit in avoiding the representation of future biotechnologies as dire threats if they were to appear instead to be mere jokes. For that reason, I have attempted to make the comedy in my stories subtle, and calculatedly bizarre as well as blithely perverse, although I often fear that the strategy might occasionally backfire, in that readers will simply not know what they are supposed to make of such stories as 'Sexual Chemistry' (Stableford, 1991), in which an unprepossessing scientific genius commissioned to produce commercial aphrodisiacs accidentally precipitates massive changes in human nature and society (as well as anticipating the discovery of Viagra); or 'The Growth of the House of Usher' (Stableford, 1991), which carefully inverts all the Gothic motifs in Poe's story by transposing its basic structure into the future, let alone a story like 'The Piebald Plumber of Haemlin' (Stableford, 2009a), which features by far the most radical transformation of human nature I have been prepared to imagine in the context of the series. I have, however, long been a believer in the principle that there is nothing as serious as comedy, especially satirical comedy, and that there are no other rhetorical devices as valuable in futuristic fiction as irony and ambiguity, simply because they create uncertainty and therefore provoke thought.

In writing his own series in direct opposition to the Frankenstein complex, Isaac Asimov was aided by the fact that his humanoid robots, being equipped with an innate morality in the form of the three laws of robotics – an authentic stroke of narrative genius – were more humane than the humans assumed to be prejudiced against them, and thus ready-made victims of human error and sin. This undoubtedly helps to account for the fact that Asimov's series is the only one of its type to have become widely popular in text form, as well as for the fact that the extant cinematic version of *I, Robot* (Proyas, 2004) is a flagrant travesty and moral betrayal of the original. Subsequent endeavours in a kindred spirit, especially those dealing with biotechnologies – such as James Blish's 'Pantropy' series (Blish, 1956) and Bruce Sterling's Shaper/Mechanist series (Sterling, 1995) – have not fared so well in reaching a large audience, in spite of their manifest brilliance as literary and imaginative endeavours.

My own endeavour has fared even less well, and I am keenly aware of the awkwardness of having been requested to write about it for an audience whose members will have had very little opportunity to discover even a few items in the series under discussion, and might well have found them puzzling and vaguely unappetising if they had the chance to do so. I hope, however, that in addition to providing a vulgar advertisement, my account of it can provide some useful insight into the difficulties of writing, publishing and finding an audience for any fiction that might potentially provide useful insights into the future of technology in general, and the future of technological human enhancement in particular.

The pioneering philosopher of science, Francis Bacon, suggested that the progress of human enlightenment is unfortunately blocked by a whole series of idols of thought, to which we cling fetishistically, the shattering of which is the price of progress. He attempted a Utopian fantasy of his own, but soon abandoned it, perhaps because he was painfully aware of its lifelessness. Employing speculative fiction in the cause of casting such idols down has never been easy, and has often required exotic strategies that had little hope of attracting widespread sympathy in the short term, but I have always believed that it is a task worth attempting, with as much ingenuity as one can muster, and with deliberate and unrepentant perversity. There is no other way in which literary work can take the side of human enhancement, rather than cravenly flattering the prejudices of the present day, which always tend toward nostalgia – but my firm belief is that literary work that merely flatters the prejudices of the present is essentially worthless, and that the cause of human enhancement is one that any rational and compassionate person ought to embrace.

References

Asimov I. (1964) *The Rest of the Robots* (Garden City NY: Doubleday).
Blish J. (1956) *The Seedling Stars* (New York: Gnome Press).
Čapek K. (1925) *The Makropoulos Secret* (Boston: Luce).
France A. (1903) *Sur la pierre blanche* (Paris: Calmann-Lévy), Charles E. Roche (tr.) *The White Stone* (London: John Lane, 1910).
Haldane J.B.S. (1923) *Daedalus; or, Science and the Future* (London: Kegan Paul, Trench & Trübner).
Huxley A. (1932) *Brave New World* (London: Chatto & Windus).
Huxley J. (1926) 'The tissue-culture king', *Amazing Stories*, August 1927, 451–9 [Originally published in *The Cornhill Magazine*, April 1926].
Mitchison N. (1975) *Solution Three* (London: Dennis Dobson).
Mitchison N. (1983) *Not by Bread Alone* (London: Marion Boyars).
Proyas A. (2004) *I, Robot* (film) (USA: 20th Century Fox).
Silverstein A. (1979) *The Conquest of Death: the Prospects for Emortality in Our Time* (New York: Macmillan).
Stableford B. (1984) *Future Man* (London: Granada).
Stableford B. (1991) *Sexual Chemistry: Sardonic Tales of the Genetic Revolution* (London: Simon & Schuster) [Includes 'And He Not Busy Being Born...', 'Cinderella's Sisters', 'The Growth of the House of Usher', 'The Invertebrate Man' and 'Sexual Chemistry'].
Stableford B. (1998) *Inherit the Earth* (New York: Tor).
Stableford B. (1999) *The Architects of Emortality* (New York: Tor).
Stableford B. (2000) *The Fountains of Youth* (New York: Tor).
Stableford B. (2001) *The Cassandra Complex* (New York: Tor).
Stableford B. (2002a) *Dark Ararat* (New York: Tor).
Stableford B. (2002b) *The Omega Expedition* (New York: Tor).
Stableford B. (2004) *Designer Genes: Tales of the Biotech Revolution* (Waterville, Maine: Five Star) [Includes 'The Age of Innocence', 'Another Branch of the Family Tree', 'The House of Mourning', 'The Invisible Worm', 'The Last Supper', 'The Milk of Human Kindness', 'The Pipes of Pan', 'Snowball in Hell' and 'What Can Chloë Want?'].
Stableford B. (2007a) *'The Cure for Love' and other Tales of the Biotech Revolution* (Borgo Press) [Includes 'Busy Dying' and 'The Road to Hell'].
Stableford B. (2007b) *'The Tree of Life' and other Tales of the Biotech Revolution* (Borgo Press) [Includes 'Hidden Agendas', 'The Home Front', 'Out of Touch', 'Rogue Terminator', 'Skin Deep', 'The Skin Trade' and 'The Tree of Life'].
Stableford B. (2009a) *'In the Flesh' and other Tales of the Biotech Revolution* (Borgo Press) [Includes 'Dr. Prospero and the Snake Lady', 'The Piebald Plumber of Haemlin' and 'The Trial'].

Stableford B. (2009b) 'Some like it hot', *Asimov's Science Fiction*, December 2009.
Stableford B. (2009c) *The Dragon Man* (Borgo Press).
Stableford B. (2009d) *'The Great Chain of Being' and other Tales of the Biotech Revolution* (Borgo Press) [Includes 'Following the Pharmers' and 'The Great Chain of Being'].
Stableford B. (2010a) *Les Fleurs du Mal* (Borgo Press).
Stableford B. (2010b) *The Undead: a Tale of the Biotech Revolution* (Borgo Press).
Stableford B. (2011a) *Nature's Shift* (Borgo Press).
Stableford B. (2011b) *Xeno's Paradox* (Borgo Press).
Stableford B. (2011c) *Zombies Don't Cry* (Borgo Press).
Stableford B. and Langford D. (1985) *The Third Millennium* (London: Sidgwick & Jackson).
Sterling B. (1995) *Schismatrix Plus* (New York: Ace).
Wilde O. (1891) *The Picture of Dorian Gray* (London: Ward Lock).

Index

AAS (Ageing Anxiety Scale), 76
accessibility
   from rehabilitation to, 120–2
   wheelchairs, 125–8, 135n10
adults, doping behaviour in, 168–9
ageing, 55, 152, *see also* disease, ageing and death
ageing anxiety, 76–7
ageism, fighting ageing and death, 77–8
AI (artificial intelligence), 139, 142–3, 183–4
Alighieri, Dante, 29
Alzheimer's disease, 21, 54, 264
Amato, Joseph, 95
amphetamines, 21, 55, 106, 110, 167, 170, 176, 231
Andrews, Lori, 32
angst, 225, 255, 257, 262
animal enhancement, 14–15, 107, 262–4
ANN (artificial neural network), 146
Annas, George, 32
anthropotechnics, 4, 39, 51–2, *see also* medicine and anthropotechnics
Aristotle, 90, 104, 141, 220
Armstrong, Lance, 235, 237–9, 251n1
Asimov, Isaac, 264, 267
athletes
   criminalising, 238–9
   doping and technical aids, 6
   doping behaviour in, 167–9
   erythropoietin, 6–7
   prevalence of prohibited substances, 168
augmentation, 2, 27, 38, 64, 65, 259, 261
authenticity, enhancement and, 25, 92–5
autonomy versus heteronomy, 154–6
Azam, Eugène, 213–14

Bacon, Francis, 10, 188, 196, 267

Bainbridge, William, 22
BCI (brain-computer interface), 144
Becker, Ernest, 75–6
Bekele, Kenenisa, 236
Bell, Thomas W., 27
Benedict XVI (Pope), 184, 199
benevolence, 97, 188
Bensaude-Vincent, Bernadette, 104
Bernal, John Desmond, 183, 190–9, 201
Besson, Colette, 236
Better than Human (Buchanan), 23
Better than Well (Elliott), 24, 39, 92
Beyond Humanity (Buchanan), 23
Beyond Therapy (President's Council on Bioethics), 38, 39–46
bioengineering, 92
biomedical enhancement, 23, 240
biotechnological enhancement, 25, 31, 47, 60, 68, 71, 79, 132, 239–40
biotechnological evolution of sport, 239–44
biotechnological illusion, 7, 14, 133
biotechnologies, 14, 21–22, 47–49, 83
   futuristic fiction and, 253–68
   history of, 262–3
Blish, James, 259, 267
Blumenberg, Hans, 185–7, 193, 200–201
BMI (brain-machine interface), 7–8, 138–48
   applications, 143–8
   bodily tools, 150–2
   CNS (central nervous system), 139, 140, 156–7n1
   computer, interface and machines, 142–3
   end effectors, 146–8, 157n2
   historical milestones, 140–3
   invasive systems, 144–8
   neuroscience, 141
   non-invasive systems, 144
   prostheses, 152–4

BMI (brain-machine
  interface) – *Continued*
  tools and human beings, 148–54
  types and places of recorded signals, 145
  types of decoding techniques and algorithms, 146
  types of operating device, 146–8
Bodnew, Ehud, 77
Bolt, Usain, 131
Bonds, Barry, 238
Bostrom, Nick, 28, 29, 34n8, 96, 98n1
Buchanan, Allen, 23
Bulwer-Lytton, Edgar, 188
Butler, Samuel, 188

Calvinist context, 11, 207–14, 221, 226n3, *see also* Strange Case of Dr Jekyll and Mr Hyde
Canguilhem, Georges, 50, 104
Čapek, Karel, 257
Carnot, Paul, 108
Carroy, Jacqueline, 213
Chambers, Dwain, 243
children, existing or future, modifying, 4, 6, 61, 63, 66, 67, 72, 92
Christianity, 183, 186–7, 200
Churchill, Winston, 189
Clayton, Lawrence, 55
cloning, 31–2
CNS (central nervous system), 139–40, 156–7n1
cocaine, 162, 206–7, 224
computers, 26, 72, 138, 144, 147, 190
Comte, Auguste, 186
Condorcet, Nicolas de, 10, 188, 199
Conrad, Peter, 113
converging technologies, 3, 22–3, 184, 197 *see also* NBIC
cosmetic enhancements, 23–4, 26, 48, 64–5, 260–1
Crawford, Robert, 76
cross-human (or collaborative) prostheses, 8, 154
cyborgs/cyborgisation, 130, 133, 135n16, 151–2, 190, 259

Daniels, Norman, 87–8

Darwin, Charles, 10, 114, 188, 189, 218
DBS (deep brain stimulation), 147
death, *see* disease, ageing and death
Defense Advanced Research Project (DARPA), 70
de Grey, Aubrey, 73–4, 81n5
De Quincey, Thomas, 221–2, 225n2
disability, 7, 8, 25, 88, 119–37, 152, 184
  repair and compensation, 119–20
  disabled sportspeople 128–32
  prostheses, 129–32
discrimination, 71–2, 77, 87
disease, ageing and death
  ageism as factor in fighting, 77–8
  fear of, 75–7
  ideal of longevity, 74–5, 81n6
  imperative to fight, 73–5
  interventions to fight, 63, 66, 67
  mass media as factor in fight, 70–1, 78
  socio-cultural influences for interventions, 73–8
doping, 7–8
  anti-doping philosophy, 235–6
  anti-doping policy, 237–8
  behaviour in sports, 8–9, 11–12
  books promoting, 53–5
  from, to doping behaviour, 163–4
  health risks of, 238
  maximising performance, 234
  moral ambiguity of, 86–7
  professional cyclists, 69
  substance consumption, 162–3
doping behaviour
  adults, 168–9
  definition of, 164–5
  identifying, 165–6
  methods for measuring, 166–7
  performance pressure and, 161–2, 171–3, 175–6
  prevalence of, 166, 170, 176n3
  studies in sports, 167–9
  studies outside sports, 169–71
  tripartite model linking performance pressure and, 174–5
  younger athletes, 169
double personality, 214, 226n8

Doyle, Arthur Conan, 189
Drexler, Eric, 183

EAPs (electrically active polymers), 147
ECoG (electrocorticogram), 144, 157*n*10
ectogenesis, 190, 193, 254
EEG (electroencephalogram), 143, 144, 157*n*12, 157*n*8
Eliot, T.S., 29
Elliott, Carl, 24, 39, 62, 68, 92
EMG (electromyogram), 144, 147, 157*n*11
emortality, 12, 253, 255–9, 262
  inheritance and, 258
  memory and, 259
end effectors, 146–8, 157*n*2
enhancement, 1–2
  augmentation, 2, 27, 38, 64, 65, 259, 261
  authenticity and, 92–5
  beyond therapy, 43, 45
  consequences of critical examination, 46–7
  continuum of property distribution, 44–5
  distinction between natural and unnatural, 45–6
  dream, dimension of, 10, 24, 52, 78
  eugenics, and, 3, 10, 87, 250
  health, 43–4
  human capacities, improvement of, 1, 3–4, 19–21, 27–8, 32–3, 44, 86, 90–1, 96, 130, 133, 230
  human dignity and, 95–8
  human nature, improvement of 3, 19, 27–32, 33, 89–92, 253, 266
  learning from, 5–9
  moral ambiguity of, 86–9
  non-specific and imprecise, 39–40
  normal *vs.* enhanced, 40–2
  pathological vs. normal, 40–2
  personal motivations for, 62–3, 67–8
  self-improvement, 3, 19, 23–7, 33
  socio-ethical issues, 60–2
  therapy-enhancement distinction, 2, 4, 13, 39, 40–2, 57, 88
  visions of future, 9–12

enhancement medicine, emergence, 230–1
enhancement techniques
  anticipating effect of modification, 113–14
  BMI (brain-machine interface), 138–48
  catalogue of, 105–6
  categories of, 107–8
  erythropoietin, 112
  genetic enhancements, 20, 31–2, 91–2, 263
  growth hormone, 2, 21, 24, 87–8, 168, 231, 237, 240
  second-stage enhancements, 184–5, 199
  see also BMI (brain-machine interface); erythropoietin (Epo)
Enlightenment, 26, 76, 90, 199
environment
  adaptation to existing, 7, 121–2, 124–7, 132, 133
  interventions for adapting to, 63, 64–5, 67
  need for social adaptation, 71–2
  pressure to adapt to fast changing, 72–3
  quest for performance, 68–71
  socio-cultural influences on motivations for adapting, 68–73
Epogen, 109, *see also* erythropoietin (Epo)
erythropoietin (Epo), 177*n*8
  history and physiology of, 108–12
  lessons from the example of, 112–13
  mutated Epo-R gene as genetic enhancement, 109–10
  as paradigm for enhancement, 110–12
  performance pressure, 173, 231
  scale of enhancement techniques, 112
evolution, Darwinian, 60, 75
extreme progress, ideology of, 10, 14, 183–5, 187, 189–91, 193, 195–200, 202
extropy, 27
Extropy Institute, 27, 28, 34*n*8

Fairhall, Neroli, 130, 135n11
Farrelly, Collin, 73–4
fiction, *see* futuristic fiction; science fiction
France, Anatole, 265
Franken, Charlotte, 196
Frankenstein (Shelley), 26
Frankenstein complex, 264, 267
Frears, Stephen, 225
Freud, Sigmund, 206, 213
Friedmann, Ted, 239
Fukuyama, Francis, 39, 91–2, 95
futuristic fiction, 9–12, 253–68
  biotechnologies, 253–68
  cosmetic enhancements, 260–1
  human skin, biotechnology, 261–62
  moral prejudices in, 265–6
  Olympics 2144 Brussels, 241–4
  Olympics of the Eight, 247–50
  performance enhancement in sport, 240–51

Gatlin, Justin, 243
Gauchet, Marcel, 187
Gebreselassie, Haile, 236
genetic engineering, 20–2, 30, 113, 259–60
genetic enhancements, 20, 31–2, 91–2, 263
  and sports, 6, 107, 109–14, 231–3
Glover, Jonathan, 21, 30, 31
Glynn, Alan, 26
Goya, Francisco, 224, 227n12
Griffith-Joyner, Florence, 234
growth hormone, 2, 21, 24, 87–8, 168, 231, 237, 240

Haldane, John B.S., 12, 34n10, 190–3, 196–9, 201, 253–5, 263
handicap, 113, 119–20, 152
handicapped persons, 87, 120
  from rehabilitation to accessibility, 120–2
happiness/well-being, interventions for, 63, 66, 67–8
Harris, John, 22, 25, 31, 88–9
Hastings Center, 20–1, 23, 34n1, 34n3

Hawthorne, Nathaniel, 219
health, 43–4
Hillis Miller, John, 221–2
Hirst, Damien, 105
horizon of meaning, 94, 98
Hottois, Gilbert, 39, 49, 139
human body
  body ownership, 8, 149, 155
  body schema, 140, 148–150, 154–5
  brain-machine interface, 148–54
  debate about enhancements, 134
  embodiment, 26, 33, 149–51, 155, 205–06, 209–10, 218, 226n9
  learning rhythm, autonomy and humanness, 154–6
  limitations of, 103–4
human capacities, *see also* enhancement
  improvement of, 1, 3–4, 19–21, 27–8, 32–3, 44, 86, 90–1, 96, 130, 133, 230
human dignity, enhancement and, 95–8
human disabilities, repair and compensation, 119–20
human enhancement, *see also* enhancement
  current visions of, 183–5
  early visions of, 188–200
  meaning of, 2–5, 19–37, 39–40
  moral ambiguity, 86–9
  modern idea of progress, 185–8
  term, 1–2, 3, 19, 32
  understanding of, 13–15
Humanity+, 28, 34n8
human nature, 19
  enhancement and, 89–92
  improvement of, 3–4, 27–32, 33
Huxley, Aldous, 141, 190, 253, 254
Huxley, Julian, 11, 29, 34n10, 190, 191, 197, 253

IAAF (International Association of Athletics Federations), 131, 239
ideal of longevity, 74–5, 81n6
identity, 3, 24–5, 26, 33, 48, 77, 89, 96–7, 129, 132, 211, 215, 226n7

ideology of extreme progress, 10, 14, 183–5, 187, 189–91, 193, 195–200, 202
modern idea of progress, 185–8
enemy of science from within, 200–202
immortality
emortality, 12, 253, 255–9, 262
ensuring, 212
invention of, 190, 194
quest for, 4, 48, 60, 66, 75
improvement, *see* enhancement
insulin-like growth factor (IGF-1), 232
invertebrate modifications, 263–4
IOC (International Olympic Committee), 236, 239, 244–7, 249
Isasi, Rosario, 32

Johnson, Ben, 243
Jonas, Hans, 156
Jones, Marion, 236–9, 243
Juvonen, Eeva, 109–10

Kass, Leon, 39, 196
Khushf, George, 184
Koch, Marita, 234
Kramer, Peter, 23–4, 95
Kratochvilova, Jamila, 234
Kurzweil, Ray, 81n4, 184, 190

La Mettrie, Julien Offray de, 199
Lamkin, Matt, 63
Landis, Floyd, 236, 237, 239
Lasch, Christopher, 24
laudanum, 206, 207, 225n2
Lee, Se-Jin, 232
Lewis, C. S., 189, 196–7
LFP (local field potential), 143, 145, 157n9
Limbo (Wolfe), 26, 151
Lippi, Giuseppe, 129
Listening to Prozac (Kramer), 23, 95
Löwith, Karl, 186

machines, interface and, 142–3, *see also* BMI (brain-machine interface)
man-machine, 133
Mannheim, Karl, 14
Mäntyranta, Eero, 6, 109–14, 236
Marx, Karl, 186

Matrix (film), 138
MBI (machine-brain interface), 147
MEA (multielectrode array), 141, 147
medicine and anthropotechnics, 48–58
deontological problem, 55–6
distinction between, 4
field and denomination, 48–50
operative distinction, 55–7
ordinary and modified, 50–2
sequence of consultation, 56–7
memory, emortality, 259
Minsky, Marvin, 183, 189
Mitchison, Naomi, 253
monsters, 11, 218, 224, 226n9, 255
Montgomery, Tim, 243
Moravec, Hans, 183, 189
More, Max, 27–9
Morris, Jan, 93

Nadal, Rafael, 238
narcissism, 24, 93, 212
National Endowment for the Humanities, 20
National Science Foundation, 3, 22
natural-artificial distinction, 6–7, 103–8, 151, 236–7, 249
arti-natural, 108, 115
and erythropoietin, 112–15
natural aptitudes or capacities, 129–3, 234
naturalistic philosophy of sport, 243, 249
natural-unnatural, 45–6, 89
Nazism, 87, 185
NBIC (nanotechnology, biotechnology, information technology, and cognitive science), 27, 60, 80n1, *see also* converging technologies
neuroscience, 141
Newmann, Vincent, 104–5
Nordmann, Alfred, 197–8
Nouvel, Pascal, 6, 103, 206, 213–14, 224–5, 226n8, 227n14

Olympic Games, 12, 106, 109–10, 129–31, 135n15, 229, 233, 236–7, 239–41, 243–7
opium, 206, 207, 225n2

ordinary and modified, 50–2
Orwell, George, 189, 196
Osbourne-Stevenson, Fanny, 213

Pantani, Marco, 234–8
Paralympics, 7, 128–31
Parens, Erik, 20, 34n1
Pearce, David, 28
pendular walking, 121, 134n2
Pereiro, Oscar, 239
Perez, Antonio, 244–7, 249
performance
  quest for, 68–71
  improvement of, 4–5, 48, 104–5, 110–11
  sexuality, performance, 70–1
performance enhancement
  fiction on, 240–51
  renewal of debate on, 229–31
  see also erythropoietin (Epo)
performance pressure, 9, 14
  doping behaviour and, 161–2, 171–3, 175–6
  tripartite model linking doping behaviour and, 174–5
personal motivations to enhance
  environment, 63, 64–5, 67
  existing or future children, 63, 66, 67
  fighting disease, ageing and death, 63, 66, 67
  happiness/well-being, 63, 66, 67–8
  identification of, 62–3, 67–8
  methods for researching, 79–80
  self-improvement, 3, 19, 23–7, 33
  socio-cultural influences on, 68–78
Persson, Ingmar, 25
PGD (prenatal genetic diagnosis), 66, 87
PIGD (pre-implantation genetic diagnosis), 87
Pistorius, Oscar, 6, 7, 25, 34n5, 106, 129–33, 134n1, 135n12, 135n15, 152
poliomyelitis, 121
Pope, Alexander, 192
posthumanism, 26
posthumanity, 32–3, 60, 91, 188, 193–4, 200, 259

precautionary principle, 114, 156
premature death, 74, 75, 81n6
principle of responsibility, 156
Principles of Extropy (Bell), 27
prostheses
  BMI and bodily tools, 150–4
  cross-human (or collaborative), 8, 154
  disabled sportspeople, 129–32
  freedom and alienation, 155–6
  learning rhythm, autonomy and humanness, 154–6
  Pistorius, 6, 7, 25
  specialised, 152–3
  supplementary, 153
  topological, 153–4
  virtualising, 153
  see also BMI (brain-machine interface)
Prozac, 23–4, 93, 95
psychostimulants, 4, 42, 48, 51, 53, 55
Pursuit of Perfection, The (Rothman and Rothman), 39, 46

Rank, Otto, 212
Reade, Winwood, 189–90, 192, 194, 197, 199
rehabilitation
  disabled sportspeople, 128–9
  from, to accessibility, 120–2
  pendular walking, 121, 134n2
Rhodes, Cecil, 189
Ricco, Riccardo, 236, 237
Riis, Bjarn, 236, 237
robotic arms, 146–8
Roco, Mihail, 22
Rothman, David, 39, 41, 49
Rothman, Sheila, 39, 41, 49
Runyan, Marla, 130

Sandel, Michael, 92
Savulescu, Julian, 25, 72
science fiction, 9, 26–7, 34n7, 34n10, 138, 151, 191, 197, 202, 253–68
Schultz, Myron, 206
secularisation, 185–8
self-improvement, 3, 19, 23–7, 33, *see also* enhancement

sexuality, performance, 70–1
Shelley, Mary, 26, 219
Shickle, Darren, 95
Sloterdijk, Peter, 39, 49, 139
Slusser, George, 197
smart drugs, 32, 48, 53–4
socio-ethical issues, human enhancement, 60–2
Souccar, Thierry, 54
sports
  biotechnological evolution, 239–40
  biotechnology, enhancement and, 231–3
  criminalising, 238–9
  debate on performance, 229–31
  doping behaviour in, 167–9
  enhancement and, 229–40
  gene therapy, 231–3
  philosophy of, 233–9
  see also disabled sportspeople
SSRIs (selective serotonin reuptake inhibitors), 93
Stableford, Brian, 12, 189, 253–68
Stapledon, Olaf, 190–1, 197
Star Trek (television series), 138
Stepanov, Yuri, 131
Sterling, Bruce, 267
Stevenson, Robert Louis, 10, 11, 26, 205–25, 226n8
stigmatisation, 71–2
Stiker, H.J., 120
Strange Case of Dr Jekyll and Mr Hyde (Stevenson), 10, 26, 205–9, 213, 220, 225, 226n3, 227n14
  Biblical background, 210–12
  Calvinistic context, 11, 207–14, 221, 226n3
  disastrous results of enhancement, 11, 219–25, 221, 226n3
  Jekyll and Hyde transformation, 205–14
  modalities of transformation, 214–19
  tragic failure, 219–25
  transcendental medicine, 11, 205, 206, 210, 215
Sweeney, H. Lee, 113, 116n16, 232

Tales of the Biotech Revolution (Stableford), 253, 260, 263, 265–6

Taylor, Charles, 93–4, 96–7, 187
technical aids, 7–8, 119–20, 122, 128–34
  repair and compensation, 119–20
  see also wheelchairs
technological interventions and objectives, examples, 64–6
Thanou, Ekaterini, 239
therapy-enhancement distinction, 2, 4, 13, 39–40, 88
TMT (Terror Management Theory), 76, 77
Tocqueville, Alexis de, 24
Tolkien, J.R.R., 196, 197
Tour de France, 229, 235–7, 239, 248
transcendental medicine, 11, 205, 206, 210, 215, see also *Strange Case of Dr. Jekyll and Mr. Hyde*
transhumanism, 3–4, 10–11, 28–30, 60, 91–2, 183–4, 188, 190–3
Tremblay, M., 121

VEPs (visual evoked potentials), 144, 157n12
Vigarello, Georges, 131
Viren, Lasse, 236

WADA (World Anti-Doping Agency), 108, 113, 163, 165, 176n2, 244
  creation of, 229–30
  ideology, 233–5
  policy, 238–9, 246
  World Anti-Doping Code, 115–16n10
Walters, LeRoy, 20, 23, 25, 34n1
Weber, Max, 96, 97
Wells, Herbert George, 34n10, 183, 189–93, 196–9, 201, 205, 219
wheelchairs, 122–8
  accessibility, 125–8
  accommodation process, 123–4
  adaptation process, 124–5
  adjustment process, 123
  creation of, 121–2
WHO (World Health Organization), 43–4, 50
Wiener, Norbert, 26
Wilde, Oscar, 227n14, 260

Wolfe, Bernard, 26, 151
WTA (World Transhumanist Association), 28, 34n8

Xeno's Paradox (Stableford), 263, 264
xenotransplantation, 256

Young, Simon, 114
yuck factor, 253–5

Zarifian, E., 50
Zatopec, Emile, 106
Zhamyatin, Yewgeny, 196
zootechnics, 14